ENTROPY AND INFORMATION OPTICS

OPTICAL ENGINEERING

Founding Editor

Brian J. Thompson

Distinguished University Professor
Professor of Optics
Provost Emeritus
University of Rochester
Rochester, New York

Editorial Board

Toshimitsu Asakura
Hokkai-Gakuen University
Sapporo, Hokkaido, Japan

Nicholas F. Borrelli
Corning, Inc.
Corning, New York

Chris Dainty
Imperial College of Science,
Technology, and Medicine
London, England

Bahram Javidi
University of Connecticut
Storrs, Connecticut

Mark Kuzyk
Washington State University
Pullman, Washington

Hiroshi Murata
The Furukawa Electric Co., Ltd.
Yokohama, Japan

Edmond J. Murphy
JDS/Uniphase
Bloomfield, Connecticut

Dennis R. Pape
Photonic Systems Inc.
Melbourne, Florida

Joseph Shamir
Technion–Israel Institute
of Technology
Hafai, Israel

David S. Weiss
Heidelberg Digital L.L.C.
Rochester, New York

1. Electron and Ion Microscopy and Microanalysis: Principles and Applications, *Lawrence E. Murr*
2. Acousto-Optic Signal Processing: Theory and Implementation, *edited by Norman J. Berg and John N. Lee*
3. Electro-Optic and Acousto-Optic Scanning and Deflection, *Milton Gottlieb, Clive L. M. Ireland, and John Martin Ley*
4. Single-Mode Fiber Optics: Principles and Applications, *Luc B. Jeunhomme*
5. Pulse Code Formats for Fiber Optical Data Communication: Basic Principles and Applications, *David J. Morris*
6. Optical Materials: An Introduction to Selection and Application, *Solomon Musikant*
7. Infrared Methods for Gaseous Measurements: Theory and Practice, *edited by Joda Wormhoudt*
8. Laser Beam Scanning: Opto-Mechanical Devices, Systems, and Data Storage Optics, *edited by Gerald F. Marshall*
9. Opto-Mechanical Systems Design, *Paul R. Yoder, Jr.*
10. Optical Fiber Splices and Connectors: Theory and Methods, *Calvin M. Miller with Stephen C. Mettler and Ian A. White*
11. Laser Spectroscopy and Its Applications, *edited by Leon J. Radziemski, Richard W. Solarz, and Jeffrey A. Paisner*
12. Infrared Optoelectronics: Devices and Applications, *William Nunley and J. Scott Bechtel*
13. Integrated Optical Circuits and Components: Design and Applications, *edited by Lynn D. Hutcheson*
14. Handbook of Molecular Lasers, *edited by Peter K. Cheo*
15. Handbook of Optical Fibers and Cables, *Hiroshi Murata*
16. Acousto-Optics, *Adrian Korpel*
17. Procedures in Applied Optics, *John Strong*
18. Handbook of Solid-State Lasers, *edited by Peter K. Cheo*
19. Optical Computing: Digital and Symbolic, *edited by Raymond Arrathoon*
20. Laser Applications in Physical Chemistry, *edited by D. K. Evans*
21. Laser-Induced Plasmas and Applications, *edited by Leon J. Radziemski and David A. Cremers*
22. Infrared Technology Fundamentals, *Irving J. Spiro and Monroe Schlessinger*
23. Single-Mode Fiber Optics: Principles and Applications, Second Edition, Revised and Expanded, *Luc B. Jeunhomme*
24. Image Analysis Applications, *edited by Rangachar Kasturi and Mohan M. Trivedi*
25. Photoconductivity: Art, Science, and Technology, *N. V. Joshi*
26. Principles of Optical Circuit Engineering, *Mark A. Mentzer*
27. Lens Design, *Milton Laikin*
28. Optical Components, Systems, and Measurement Techniques, *Rajpal S. Sirohi and M. P. Kothiyal*
29. Electron and Ion Microscopy and Microanalysis: Principles and Applications, Second Edition, Revised and Expanded, *Lawrence E. Murr*

30. Handbook of Infrared Optical Materials, *edited by Paul Klocek*
31. Optical Scanning, *edited by Gerald F. Marshall*
32. Polymers for Lightwave and Integrated Optics: Technology and Applications, *edited by Lawrence A. Hornak*
33. Electro-Optical Displays, *edited by Mohammad A. Karim*
34. Mathematical Morphology in Image Processing, *edited by Edward R. Dougherty*
35. Opto-Mechanical Systems Design: Second Edition, Revised and Expanded, *Paul R. Yoder, Jr.*
36. Polarized Light: Fundamentals and Applications, *Edward Collett*
37. Rare Earth Doped Fiber Lasers and Amplifiers, *edited by Michel J. F. Digonnet*
38. Speckle Metrology, *edited by Rajpal S. Sirohi*
39. Organic Photoreceptors for Imaging Systems, *Paul M. Borsenberger and David S. Weiss*
40. Photonic Switching and Interconnects, *edited by Abdellatif Marrakchi*
41. Design and Fabrication of Acousto-Optic Devices, *edited by Akis P. Goutzoulis and Dennis R. Pape*
42. Digital Image Processing Methods, *edited by Edward R. Dougherty*
43. Visual Science and Engineering: Models and Applications, *edited by D. H. Kelly*
44. Handbook of Lens Design, *Daniel Malacara and Zacarias Malacara*
45. Photonic Devices and Systems, *edited by Robert G. Hunsperger*
46. Infrared Technology Fundamentals: Second Edition, Revised and Expanded, *edited by Monroe Schlessinger*
47. Spatial Light Modulator Technology: Materials, Devices, and Applications, *edited by Uzi Efron*
48. Lens Design: Second Edition, Revised and Expanded, *Milton Laikin*
49. Thin Films for Optical Systems, *edited by François R. Flory*
50. Tunable Laser Applications, *edited by F. J. Duarte*
51. Acousto-Optic Signal Processing: Theory and Implementation, Second Edition, *edited by Norman J. Berg and John M. Pellegrino*
52. Handbook of Nonlinear Optics, *Richard L. Sutherland*
53. Handbook of Optical Fibers and Cables: Second Edition, *Hiroshi Murata*
54. Optical Storage and Retrieval: Memory, Neural Networks, and Fractals, *edited by Francis T. S. Yu and Suganda Jutamulia*
55. Devices for Optoelectronics, *Wallace B. Leigh*
56. Practical Design and Production of Optical Thin Films, *Ronald R. Willey*
57. Acousto-Optics: Second Edition, *Adrian Korpel*
58. Diffraction Gratings and Applications, *Erwin G. Loewen and Evgeny Popov*
59. Organic Photoreceptors for Xerography, *Paul M. Borsenberger and David S. Weiss*
60. Characterization Techniques and Tabulations for Organic Nonlinear Optical Materials, *edited by Mark Kuzyk and Carl Dirk*

61. Interferogram Analysis for Optical Testing, *Daniel Malacara, Manuel Servín, and Zacarias Malacara*
62. Computational Modeling of Vision: The Role of Combination, *William R. Uttal, Ramakrishna Kakarala, Sriram Dayanand, Thomas Shepherd, Jagadeesh Kalki, Charles F. Lunskis, Jr., and Ning Liu*
63. Microoptics Technology: Fabrication and Applications of Lens Arrays and Devices, *Nicholas F. Borrelli*
64. Visual Information Representation, Communication, and Image Processing, *Chang Wen Chen and Ya-Qin Zhang*
65. Optical Methods of Measurement: Wholefield Techniques, *Rajpal S. Sirohi and Fook Siong Chau*
66. Integrated Optical Circuits and Components: Design and Applications, *edited by Edmond J. Murphy*
67. Adaptive Optics Engineering Handbook, *edited by Robert K. Tyson*
68. Entropy and Information Optics, *Francis T. S. Yu*

Additional Volumes in Preparation

Computational Methods for Electromagnetic and Optical Systems, *John M. Jarem and Partha P. Banerjee*

Laser Beam Shaping: Theory and Techniques, *edited by Fred M. Dickey and Scott C. Holswade*

ENTROPY AND INFORMATION OPTICS

FRANCIS T. S. YU
The Pennsylvania State University
University Park, Pennsylvania

MARCEL DEKKER, INC. NEW YORK • BASEL

ISBN: 0-8247-0363-4

This book is printed on acid-free paper.

Headquarters
Marcel Dekker, Inc.
270 Madison Avenue, New York, NY 10016
tel: 212-696-9000; fax: 212-685-4540

Eastern Hemisphere Distribution
Marcel Dekker AG
Hutgasse 4, Postfach 812, CH-4001 Basel, Switzerland
tel: 41-61-261-8482; fax: 41-61-261-8896

World Wide Web
http://www.dekker.com

The publisher offers discounts on this book when ordered in bulk quantities. For more information, write to Special Sales/Professional Marketing at the headquarters address above.

Copyright © 2000 by Marcel Dekker, Inc. All Rights Reserved.

Neither this book nor any part may be reproduced or transmitted in any form or by any means, electronic or mechanical, including photocopying, microfilming, and recording, or by any information storage and retrieval system, without permission in writing from the publisher.

Current printing (last digit):
10 9 8 7 6 5 4 3 2 1

PRINTED IN THE UNITED STATES OF AMERICA

From the Series Editor

Fundamentally optical beams and optical systems transmit and analyze information. The information can be analog or digital. It can be three-dimensional, two-dimensional, or one-dimensional. It can be in the traditional form of an image or information that is coded and/or compressed. The light beam carrying the information can be incoherent, coherent, or even partially coherent.

In the early days of this important field, the concepts of communication theory had a major impact on our understanding and our descriptions of optical systems. The initial impetus was to deal with images and image quality. Concepts of impulse responses and transfer functions caused considerable rethinking about the design and evaluation of optical systems. Resolution criteria were only the beginning; "fidelity," "fidelity defect," "relative structural content," and "correlation quantity" were concepts introduced by E. H. Linfoot in 1964. Formal definitions of entropy and information content were to follow and the field continues to expand, driven by the explosion of high-speed, high-data-rate and high-capacity communication systems.

This volume discusses the fundamentals and the applications of entropy and information optics by means of a sampling of topics in this field, including image restoration, wavelet transforms, pattern recognition, computing, and fiber-optic communication.

Brian J. Thompson

Preface

> Light is one of the most important information carriers in space.
>
> One cannot get something from nothing, even by observation.

The discovery of the laser in the 1960s prompted the building of new optical communication and processing systems. The impact of fiber-optic communication and optical signal processing provided unique evidence of the relationship between optics and information theory. As we are all aware, light not only is the main source of energy that supports life but is also a very important carrier of information. Therefore, my objective here is to describe the profound relationship between entropy and information optics. My earlier book, *Optics and Information Theory* (Wiley, 1976), has been read and used as a text by numerous universities and by engineers in the United States and abroad. Using that work as a base, I have incorporated in this book the vast amount of new developments in the field.

The contents of this book, in part, have been used as course notes in my classes taught at The Pennsylvania State University. The materials were found to be both stimulating and enlightening. They should provide a deeper appreciation of optics for readers. Nevertheless, a book of this form is designed not to cover the vast domain of entropy and information optics but to focus on a few areas that are of particular interest.

The relationship between entropy information and optics has provided the basic impetus for research on and development of high-speed, high-data-rate, and high-capacity communication systems. This trend started some years ago and will continue to become more widespread in years to come. The reason for this success may be deduced from the imaginative relationship between entropy information and optics that is described in this book.

Finally, I would like to express my sincere appreciation to my colleagues for their enthusiastic encouragement: without their support this work would not have been completed.

Francis T. S. Yu

Contents

From the Series Editor Brian J. Thompson		*iii*
Preface		*v*

1. Introduction to Information Transmission   **1**
    1.1   Information Measure   4
    1.2   Entropy Information   7
    1.3   Communication Channels   15
    1.4   Memoryless Discrete Channels   16
    1.5   Continuous Channels with Additive Noise   22
    1.6   Summary and Remarks   31
    References   33

2. Diffraction and Signal Analysis   **35**
    2.1   Introduction to Diffraction   36
    2.2   Fresnel–Kirchhoff Theory   39
    2.3   Linear Systems and Fourier Analysis   41
    2.4   Finite Bandwidth Analysis   44
    2.5   Degrees of Freedom of a Signal   49
    2.6   Gabor's Information Cell   52
    2.7   Signal Detection   55
    2.8   Statistical Signal Detection   57
    2.9   Signal Recovering   61
    2.10   Signal Ambiguity   63
    2.11   Wigner Signal Representation   66
    2.12   Fourier Transform Properties of Lenses   70
    References   73

3. Optical Spatial Channel and Encoding Principles   **75**
    3.1   Optical Spatial Communication Channel   75
    3.2   Optical Message in Spatial Coding   80

	3.3	Optical Channel with Resolution Cells of Different Sizes	84
	3.4	Matching a Code with a Spatial Channel	89
		References	93
4.	**Entropy and Information**	**95**
	4.1	Fundamental Laws of Thermodynamics	95
	4.2	Physical Entropy and Information	99
	4.3	Trading Entropy with Information	100
	4.4	Typical Examples	102
	4.5	Remarks	106
		References	107
5.	**Demon Exorcist and Cost of Entropy**	**109**
	5.1	Perpetual Motion Machine	109
	5.2	Maxwell's Demon	111
	5.3	Information and Demon Exorcist	113
	5.4	Demon Exorcist, A Revisit	118
	5.5	Szilard's Demon	121
	5.6	Diffraction-Limited Demon	124
	5.7	Minimum Cost of Entropy	126
	5.8	Gabor's Perpetuum Mobile of the Second Kind	129
		References	131
6.	**Observation and Information**	**133**
	6.1	Observation with Radiation	135
	6.2	Simultaneous Observations	139
	6.3	Observation and Information	142
	6.4	Accuracy and Reliability in Observations	145
	6.5	Observation by Interference and by Microscope	153
	6.6	Uncertainty and Observation	159
	6.7	Remarks	162
		References	163
7.	**Image Restoration and Information**	**165**
	7.1	Image Restoration	165
	7.2	Uncertainty and Image Restoration	171
	7.3	Resolving Power and Information	175
	7.4	Coherent and Digital Image Enhancement	178
	7.5	Information Leakage through a Passive Channel	180
	7.6	Restoration of Blurred Images	186
		References	192

8.	**Quantum Effect on Information Transmission**	**195**
	8.1 Problem Formulation and Entropy Consideration	196
	8.2 Capacity of a Photon Channel	198
	8.3 An Informational Theoristic Approach	203
	8.4 Narrow-Band Photon Channel	206
	8.5 Optimum Signal Power Distribution, A Special Case	212
	References	215
9.	**Coherence Theory of Optics**	**217**
	9.1 Aspects of Coherence	217
	9.2 Spatial and Temporal Coherence	221
	9.3 Coherent and Incoherent Processing	224
	9.4 Exploitation of Coherence	226
	9.5 Remarks	232
	References	232
10.	**Wavelet Transforms with Optics**	**233**
	10.1 Aspects of Wavelet Transform	233
	10.2 Fourier Domain Processing	234
	10.3 Wavelet Transform	239
	10.4 Optical Implementations	242
	10.5 Simulations	246
	10.6 Remarks	249
	References	249
11.	**Pattern Recognition with Optics**	**251**
	11.1 Optical Correlators	251
	11.2 Optical-Disk-Based Correlator	255
	11.3 Photorefractive-Based Correlator	257
	11.4 Optical Neural Networks	259
	11.5 Composite Filters	263
	11.6 Remarks	269
	References	269
12.	**Computing with Optics**	**271**
	12.1 Logic-Based Computing	271
	12.2 Optical-Interconnects and Shuffling	275
	12.3 Matrix-Vector Multiplication	278
	12.4 Systolic Processor	279
	12.5 Matrix-Matrix Processing	282
	12.6 Expert System and Artificial Intelligence	284
	12.7 Remarks	292
	References	293

13.	Communication with Fiber Optics	**295**
	13.1 Aspects of Fiber-Optic Communication	295
	13.2 Optical Fiber Structures	301
	13.3 Fiber-Optic Transmission	303
	13.4 Types of Optical Fibers	309
	13.5 Fiber-Optic Communications	312
	13.6 Remarks	315
	References	317

Appendix A	*Linear Difference Equation with Constant Coefficients*	*319*
Appendix B	*Solution of the* a priori *Probabilities of Eqs. (5.37) and (5.38)*	*321*
Appendix C	*Probability Energy Distribution*	*323*

Index   *325*

1
Introduction to Information Transmission

In the physical world, light is not only part of the mainstream of energy that supports life; it also provides us with important sources of information. One can easily imagine that without light present civilization would never have emerged. Furthermore, humans are equipped with a pair of exceptionally good, although not perfect, eyes. With the combination of an intelligent brain and remarkable eyes, humans were able to advance themselves above the rest of the animals in the world. It is undoubtedly true that if humans had not been equipped with eyes, they would not have evolved into their present form. In the presence of light, humans are able to search for the food they need and the art they enjoy, and to explore the unknown. Thus light, or rather *optics*, has provided us with a very useful source of information whose application can range from very abstract artistic to very sophisticated scientific uses.

The purpose of this text is to discuss the relationship between optics and information transmission. However, it is emphasized that it is not our intention to consider the whole field of optics and information, but rather to center on an area that is important and interesting to our readers.

Prior to going into a detailed discussion of optics and information, we devote this first chapter to the fundamentals of information transmission. However, it is noted that *entropy information* was not originated by optical physicists, but rather by a group of mathematically oriented electrical engineers whose original interest was centered on electrical communication. Nevertheless, from the very beginning of the discovery of entropy information, interest in the application has never totally been absent from the optical standpoint. As a result of the recent advances in modern information optics and optical communication, the relationship between optics and entropy information has grown more rapidly than ever.

Although everyone seems to know the word information, a fundamental theoristic concept may not be the case. Let us now define the meaning of information. Actually, information may be defined in relation to several different disciplines. In fact, information may be defined according to its applications but with the identical mathematical formalism as developed in the next few sections. From the viewpoint of pure mathematics, information theory is basically a *probabilistic concept*. We see in Sec. 1.1 that without probability there would be no information theory. But, from a physicist's point of view, information theory is essentially an *entropy theory*. In Chap. 4, we see that without the fundamental relationship between physical entropy and information entropy, information theory would have no useful application in physical science. From a communication engineer's standpoint, information theory can be considered an *uncertainty theory*. For example, the more uncertainty there is about a message we have received, the greater the amount of information the message contained.

Since it is not our intention to define information for all fields of interest, we quickly summarize: The beauty and greatness of entropy of information is its' application to all fields of science. Application can range from the very abstract (e.g., economy, music, biology, psychology) to very sophisticated hardcore scientific researches. However, in our present introductory version, we consider the concept of information from a practical communication standpoint. For example, from the information theory viewpoint, a perfect liar is as good an informant as a perfectly honest person, provided of course that we have the *a priori* knowledge that the person is a perfect liar or perfectly honest. One should be cautious not to conclude that if one cannot be an honest person, one should be a liar. For, as we may all agree, the most successful crook is the one that does not look like one. Thus we see that information theory is a guessing game, and is in fact a *game theory*.

In general, an information-transmission system can be represented by a block diagram, as shown in Fig. 1.1. For example, in simple optical communication, we have a message (an information source) shown by means of written characters, for example, Chinese, English, French, German. Then we select suitable written characters (a code) appropriate to our communication. After the characters are selected and written on a piece of paper, the information still cannot be transmitted until the paper is illuminated by visible light (the transmitter), which obviously acts as an information carrier. When light reflected from the written characters arrives at your eyes (the receiver), a proper decoding (translating) process takes place, that is, character recognition (decoding) by the user (your mind). Thus, from this simple example, we can see that a suitable encoding process may not be adequate unless a suitable decoding process also takes place. For instance, if I show you a Chinese newspaper you might not be able to decode the language, even if the optical channel is assumed to be perfect (i.e.,

Introduction to Information Transmission

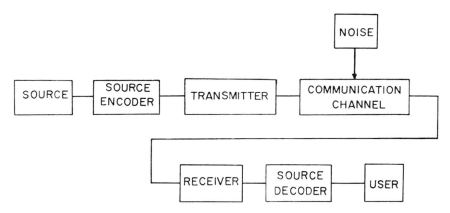

Fig. 1.1 Block diagram of a communication system.

noiseless). This is because a suitable decoding process requires *a priori* knowledge of the encoding scheme (i.e., appropriate information storage), for example, *a priori* knowledge of the Chinese characters. Thus the decoding process can also be called a *recognition process*.

Information theory is a broad subject which can not be fully discussed in a few sections. Although we only investigate the theory in an introductory manner, our discussion in the next few sections provides a very useful application of entropy information to optics. Readers who are interested in a rigorous treatment of information theory are referred to the classic papers by Shannon [1–3] and the text by Fano [4].

Information theory has two general orientations: one developed by Wiener [5, 6], and the other by Shannon [1–3]. Although both Wiener and Shannon share a common probabilistic basis, there is a basic distinction between them.

The significance of Wiener's work is that, if a signal (information) is corrupted by some physical means (e.g., noise, nonlinear distortion), it may be possible to recover the signal from the corrupted one. It is for this purpose that Wiener develops the theories of correlation detection, optimum prediction, matched filtering, and so on. However, Shannon's work is carried a step further. He shows that the signal can be optimally transferred provided it is properly encoded. That is, the signal to be transferred can be processed before and after transmission. In the encoding process, he shows that it is possible to combat the disturbances in the communication channel to a certain extent. Then, by a proper decoding process, the signal can be recovered optimally. To do this, Shannon develops the theories of information measure, channel capacity, coding processes, and so on. The major interest in Shannon's theory is efficient utilization of the communication channel.

A fundamental theorem proposed by Shannon can be considered the most surprising result of his work. The theorem can be stated approximately as follows. Given a stationary finite-memory information channel having a channel capacity C, if the binary information transmission rate R of the message is smaller than C, there exists channel encoding and decoding processes for which the probability of error in information transmission per digit can be made arbitrarily small. Conversely, if the information transmission rate R is larger than C, there exists no encoding and decoding processes with this property; that is, the probability of error in information transmission cannot be made arbitrarily small. In other words, the presence of random disturbances in a communication channel does not, by itself, limit transmission accuracy. Rather, it limits the transmission rate for which arbitrarily high transmission accuracy can be accomplished.

In summarizing this brief introduction to information transmission, we point out again the distinction between the viewpoints of Wiener and of Shannon. Wiener assumes in effect that the signal in question can be processed after it has been corrupted by noise. Shannon suggests that the signal can be processed both before and after its transmission through the communication channel. However, the main objectives of these two branches of information transmission are basically the same, namely, faithful reproduction of the original signal.

1.1 INFORMATION MEASURE

We have in the preceding discussed a general concept of information transmission. In this section, we discuss this subject in more detail. Our first objective is to define a measure of information, which is vitally important in the development of modern information theory. We first consider discrete input and discrete output message ensembles as applied to a communication channel, as shown in Fig. 1.2. We denote the sets of input and output ensembles $A = \{a_i\}$ and $B = \{b_j\}$, respectively, $i = 1, 2, \ldots, M$, and $j = 1, 2, \ldots, N$. It is noted that AB forms a *discrete product space*.

Let us assume that a_i is an input event to the information channel, and b_j is the corresponding output event. Now we would like to define a measure of information in which the received event b_j specifies a_i. In other words, we would like to define a measure of the amount of information provided by the output event b_j *about* the corresponding input event a_i. We see that the transmission of a_i through the communication channel causes a change

Introduction to Information Transmission

Fig. 1.2 An input-output communication channel.

in the probability of a_i, from an *a priori* $P(a_i)$ to an a posteriori $P(a_i/b_j)$. In measuring this change, we take the logarithmic ratio of these probabilities. It turns out to be appropriate for the definition of information measure. Thus the amount of information provided by the output event b_j about the input event a_i can be defined as

$$I(a_i; b_j) \triangleq \log_2 \frac{P(a_i/b_j)}{P(a_i)} \qquad \text{bits} \qquad (1.1)$$

It is noted that the base of the logarithm can be a value other than 2. However, the base 2 is the most commonly used in information theory. Therefore we adopt this base value of 2 for use in this text. Other base values are also frequently used, for example, $\log_{10}$ and $\ln = \log_e$. The corresponding units of information measure of these different bases are hartleys and nats. The *hartley* is named for R. V. Hartley, who first suggested the use of a logarithmic measure of information [7], and *nat* is an abbreviation for *natural unit*. Bit, used in Eq. (1.1), is a contraction of *binary unit*.

We see that Eq. (1.1) possesses a symmetric property with respect to input event a_i and output event b_j:

$$I(a_i; b_j) = I(b_j; a_i) \qquad (1.2)$$

This symmetric property of information measure can be easily shown:

$$\log_2 \frac{P(a_i, b_j)}{P(b_j)P(a_i)} = \log_2 \frac{P(b_j/a_i)}{P(b_j)}$$

According to Eq. (1.2), the amount of information provided by event b_j about event a_i is the same as that provided by a_i about b_j. Thus Eq. (1.1) is a measure defined by Shannon as *mutual information* or amount of information transferred between event a_i and event b_j.

It is clear that, if the input and output events are *statistically independent*, that is, if $P(a_i, b_j) = P(a_i)P(b_j)$, then $I(a_i; b_j) = 0$.

Furthermore, if $I(a_i; b_j) > 0$, then $P(a_i, b_j) > P(a_i)P(b_j)$, that is, there is a higher joint probability of a_i and b_j. However, if $I(a_i; b_j) < 0$, then $P(a_i, b_j) < P(a_i)P(b_j)$, that is, there is a lower joint probability of a_i and b_j.

As a result of the conditional probabilities $P(a_i/b_j) \leq 1$, and $P(b_j/a_i) \leq 1$, we see that

$$I(a_i; b_j) \leq I(a_i) \tag{1.3}$$

and

$$I(a_i; b_j) \leq I(b_j) \tag{1.4}$$

where

$$I(a_i) \triangleq -\log_2 P(a_i) \tag{1.5}$$

$$I(b_j) \triangleq -\log_2 P(b_j) \tag{1.6}$$

$I(a_i)$ and $I(b_j)$ are defined as the respective *input* and *output self-information* of event a_i and event b_j. In other words, $I(a_i)$ and $I(b_j)$ represent the amount of information provided at the input and output of the information channel of event a_i and event b_j, respectively. It follows that the mutual information of event a_i and event b_j is equal to the self-information of event a_i if and only if $P(a_i/b_j) = 1$; that is,

$$I(a_i; b_j) = I(a_i) \tag{1.7}$$

It is noted that, if Eq. (1.7) is true for all i, that is, the input ensemble, then the communication channel is *noiseless*. However, if $P(b_j/a_i) = 1$, then

$$I(a_i; b_j) = I(b_j) \tag{1.8}$$

If Eq. (1.8) is true for all the output ensemble, then the information channel is *deterministic*.

It is emphasized that the definition of measure of information can be extended to higher product spaces. For example, we can define the mutual information for a product ensemble ABC:

$$I(a_i; b_j/c_k) \triangleq \log_2 \frac{P(a_i/b_j c_k)}{P(a_i/c_k)} \tag{1.9}$$

Similarly, one can show that

$$I(a_i; b_j/c_k) = I(b_j; a_i/c_k) \tag{1.10}$$

$$I(a_i; b_j/c_k) \leq I(a_i/c_k) \tag{1.11}$$

and

$$I(a_i; b_j/c_k) \leq I(b_j/c_k) \tag{1.12}$$

where

$$I(a_i/c_k) \triangleq -\log_2 P(a_i/c_k) \tag{1.13}$$

Introduction to Information Transmission 7

and

$$I(b_j/c_k) \triangleq -\log_2 P(b_j/c_k) \tag{1.14}$$

represent the *conditional self-information*:
Furthermore, from Eq. (1.1) we see that

$$I(a_i; b_j) = I(a_i) - I(a_i/b_j) \tag{1.15}$$

and

$$I(a_i; b_j) = I(b_j) - I(b_j/a_i) \tag{1.16}$$

From the definition of

$$I(a_i b_j) \triangleq -\log_2 P(a_i, b_j) \tag{1.17}$$

the self-information of the point (a_i, b_j) of the product ensemble AB, one can show that

$$I(a_i; b_j) = I(a_i) + I(b_j) - I(a_i b_j) \tag{1.18}$$

Conversely,

$$I(a_i b_j) = I(a_i) + I(b_j) - I(a_i; b_j) \tag{1.19}$$

In concluding this section, we point out that, for the mutual information $I(a_i; b_j)$ (i.e., the amount of information transferred through the channel) there exists an upper bound, $I(a_i)$ or $I(b_j)$, whichever comes first. If the information channel is noiseless, then the mutual information $I(a_i; b_j)$ is equal to $I(a_i)$, the input self-information of a_i. However, if the information channel is deterministic, then the mutual information is equal to $I(b_j)$, the output self-information of b_j. Moreover, if the input-output of the information channel is statistically independent, then no information can be transferred. It is also noted that, when the joint probability $P(a_i; b_j) < P(a_i)P(b_j)$, then $I(a_i; b_j)$ is negative, that is, the information provided by event b_j about event a_i further deteriorates, as compared with the statistically independent case. Finally, it is clear that the definition of the measure of information can also be applied to a higher product ensemble, namely, $ABC \cdots$ produce space.

1.2 ENTROPY INFORMATION

In Sec. 1.1 we defined a measure of information. We saw that information theory is indeed a branch of probability theory.

In this section, we consider the measure of information as a random variable, that is, information measure as a random event. Thus the measure of information can be described by a probability distribution $P(I)$, where I is the self-, conditional, or mutual information.

Since the measure of information is usually characterized by an ensemble average, the average amount of information provided can be obtained by the ensemble average

$$E[I] = \sum_I IP(I) \tag{1.20}$$

where E denotes the ensemble average, and the summation is over all I.

If the self-information a_i in Eq. (1.5) is used in Eq. (1.20), then the average amount of self-information provided by the input ensemble A is

$$E[I(a_i)] = \sum_I IP(I) = \sum_{i=1}^M P(a_i) I(a_i) \tag{1.21}$$

where $I(a_i) = -\log_2 P(a_i)$.

For convenience in notation, we drop the subscript i; thus Eq. (1.21) can be written

$$I(A) \triangleq -\sum_A P(a) \log_2 P(a) \triangleq H(A) \tag{1.22}$$

where the summation is over the input ensemble A.

Similarly, the average amount of self-information provided at the output end of the information channel can be written

$$I(B) \triangleq -\sum_B P(b) \log_2 P(b) \triangleq H(B) \tag{1.23}$$

As a matter of fact, Eqs. (1.22) and (1.23) are the starting points of Shannon's [1–3] information theory. These two equations are in essentially the same form as the *entropy equation* in statistical thermodynamics. Because of the identical form of the entropy expression, $H(A)$ and $H(B)$ are frequently used to describe *information entropy*. Moreover, we see in the next few chapters that Eqs. (1.22) and (1.23) are not just mathematically similar to the entropy equation, but that they represent a profound relationship between science and information theory [8–10], as well as between optics and information theory [11,12].

It is noted that entropy H, from the communication theory point of view, is mainly a measure of *uncertainty*. However, from the statistical thermodynamic point of view, entropy H is a measure of *disorder*.

In addition, from Eqs. (1.22) and (1.23), we see that

$$H(A) \geq 0 \tag{1.24}$$

where $P(a)$ is always a positive quantity. The equality of Eq. (1.24) holds if $P(a) = 1$ or $P(a) = 0$. Thus we can conclude that

$$H(A) \leq \log_2 M \tag{1.25}$$

where M is the number of different events in the set of input events A, that is, $A = \{a_i\}, i = 1, 2, \ldots, M$. We see that the equality of Eq. (1.25) holds if and only if $P(a) = 1/M$, that is, if there is *equiprobability* of all the input events.

In order to prove the inequality of Eq. (1.25), we use the well-known inequality

$$\ln u \leq u - 1 \tag{1.26}$$

Let us now consider that

$$\begin{aligned} H(A) - \log_2 M &= -\sum_A p(a) \log_2 p(a) - \sum_A p(a) \log_2 M \\ &= \sum_A p(a) \log_2 \frac{1}{Mp(a)} \end{aligned} \tag{1.27}$$

By the use of Eq. (1.26), one can show that

$$H(A) - \log_2 M \leq \sum_A \left[\frac{1}{M} - p(a) \right] \log_2 e = 0 \tag{1.28}$$

Thus we have proved that the equality of Eq. (1.25) holds if and only if the input ensemble is equiprobable, $p(a) = 1/M$. We see that the entropy $H(A)$ is maximum when the probability distribution a is equiprobable. Under the maximization condition of $H(A)$, the amount of information provided is the *information capacity* of A.

To show the behavior of $H(A)$, we describe a simple example for the case of $M = 2$, that is, for a *binary source*. Then the entropy equation (1.22) can be written as

$$H(p) = -p \log_2 p - (1-p) \log_2 (1-p) \tag{1.29}$$

where p is the probability of one of the events.

From Eq. (1.29) we see that $H(p)$ is maximum if and only if $p = \frac{1}{2}$. Moreover, the variation in entropy as a function of p is plotted in Fig. 1.3, in which we see that $H(p)$ is a symmetric function, having a maximum value of 1 bit at $p = \frac{1}{2}$.

Similarly, one can extend this concept of ensemble average to the conditional self-information:

$$I(B/A) \triangleq -\sum_B \sum_A p(a, b) \log_2 p(b/a) \triangleq H(B/A) \tag{1.30}$$

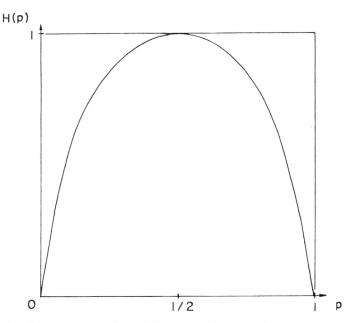

Fig. 1.3 The variation of $H(p)$ as a function of p.

We define $H(B/A)$ as the conditional entropy of B given A. Thus the entropy of the product ensemble AB can also be written

$$H(AB) = -\sum_A \sum_b p(a,b) \log_2 p(a,b) \tag{1.31}$$

where $p(a,b)$ is the joint probability of events a and b.

From the entropy equations (1.22) and (1.30), we have the relation

$$H(AB) = H(A) + H(B/A) \tag{1.32}$$

Similarly, we have

$$H(AB) = H(B) + H(A/B) \tag{1.33}$$

where

$$H(A/B) = -\sum_A \sum_B p(a,b) \log_2 p(a/b) \tag{1.34}$$

From the relationships of Eq. (1.26), we also show that

$$H(B/A) \leq H(B) \tag{1.35}$$

and
$$H(A/B) \leq H(A) \tag{1.36}$$
where the equalities hold if and only if a and b are statistically independent.

Furthermore, Eqs. (1.35) and (1.36) can be extended to a higher product ensemble space. For example, with a triple product space ABC, we have the conditional entropy relation

$$H(C/AB) \leq H(C/B) \tag{1.37}$$

in which the equality holds if and only if c is statistically independent of a for any given b, that is, if $p(c/ab) = p(c/b)$.

It is noted that extension of the conditional entropy relationship to a higher product ensemble, for example, source encoding is of considerable importance. Since the conditional entropy is the average amount of information provided by the successive events, it cannot be increased by making the successive events dependent on the preceding ones. Thus we see that the information capacity of an encoding alphabet cannot be made maximum if the successive events are interdependent. Therefore, the entropy of a message ensemble places a lower limit on the average number of coding digits per code word:

$$\bar{n} \geq \frac{H(A)}{\log_2 D} \tag{1.38}$$

where $\bar{n}$ is the average number of coded digits and D is the number of the *coding alphabet*; for example, for binary coding, the number of the coding alphabet is 2. It is emphasized that the lower limit of Eq. (1.38) can be approached as closely as we desire for encoding sufficiently long sequences of independent messages. However, long sequences of messages also involve a more complex coding procedure.

We now turn our attention to defining the *average mutual information*. We consider first the conditional average mutual information:

$$I(A;\, b) \triangleq \sum_A p(a/b) I(a;\, b) \tag{1.39}$$

where

$$I(a;\, b) = \log_2 \frac{p(a/b)}{p(a)}$$

Although the mutual information of an event a and an event b can be negative, $I(a; b) < 0$, the average conditional mutual information can never be negative:

$$I(A;\, b) \geq 0 \tag{1.40}$$

with the equality holding if and only if events A are statistically independent of b, that is, $p(a/b) = p(a)$, for all a.

By taking the ensemble average of Eq. (1.39), the average mutual information can be defined:

$$I(A; B) \triangleq \sum_B p(b) I(A; b) \tag{1.41}$$

Equation (1.41) can be written

$$\begin{aligned} I(A; B) &\triangleq \sum_B \sum_A p(a, b) \log_2 \frac{p(a/b)}{p(a)} \\ &= \sum_A \sum_B p(a, b) I(a, b) \end{aligned} \tag{1.42}$$

Again one can show that

$$I(A; B) \geq 0 \tag{1.43}$$

The equality holds for Eq. (1.43) if and only if a and b are statistically independent. Moreover, from the symmetric property of $I(a, b)$ [Eq. (1.2)], it can be easily shown that

$$I(A; B) = I(B; A) \tag{1.44}$$

where

$$I(B; A) \triangleq \sum_A \sum_B p(a, b) \log_2 \frac{p(b/a)}{p(b)} \tag{1.45}$$

Furthermore, from Eqs. (1.3) and (1.4), one can show that

$$I(A; B) \leq H(A) = I(A) \tag{1.46}$$

and

$$I(A; B) \leq H(B) = I(B) \tag{1.47}$$

This says that the mutual information (the amount of information transfer) cannot be greater than the entropy (the amount of information provided) at the input or the output ends of the information channel, whichever comes first. We see that, if the equality holds for Eq. (1.46), then the channel is noiseless; however, if the equality holds for Eq. (1.47), then the channel is deterministic.

From the entropy equation (1.31), we can show that

$$H(AB) = H(A) + H(B) - I(A; B) \tag{1.48}$$

From the relationship of Eq. (1.48) and the conditional entropy of Eqs. (1.32) and (1.33), we have

$$I(A; B) = H(A) - H(A/B) \qquad (1.49)$$

and

$$I(A; B) = H(B) - H(B/A) \qquad (1.50)$$

Equations (1.49) and (1.50) are of interest to us in determination of the mutual information (the amount of information transfer). For example, if $H(A)$ is considered the average amount of information provided at the input end of the channel, then $H(A/B)$ is the average amount of *information loss* (e.g., due to noise) in the channel. It is noted that the conditional entropy $H(A/B)$ is usually regarded as the *equivocation* of the channel. However, if $H(B)$ can be considered the average amount of information received at the output end of the channel, then $H(B/A)$ is the average amount of information needed to specify the noise disturbance in the channel. Thus $H(B/A)$ may be referred to as the *noise entropy* of the channel. Since the concept of mutual information can be extended to a higher product ensemble, we can show that [13]

$$I(A; BC) = I(A; B) + I(A; C/B) \qquad (1.51)$$

and

$$I(BC; A) = I(B; A) + I(C; A/B) \qquad (1.52)$$

By the symmetric property of the mutual information, we define *triple mutual information*:

$$\begin{aligned} I(a; b; c) &\triangleq I(a; b) - I(a; b/c) = I(a; c) - I(a; c/b) \\ &= I(b; c) - I(b; c/a) \end{aligned} \qquad (1.53)$$

Thus we have

$$\begin{aligned} I(A; B; C) &\triangleq \sum_A \sum_B \sum_C p(a, b, c) I(a; b; c) \\ &= I(A; B) - I(A; B/C) \\ &= I(A; C) - I(A; C/B) \\ &= I(B; C) - I(B; C/A) \end{aligned} \qquad (1.54)$$

In view of Eq. (1.54), it is noted that $I(A; B; C)$ can be positive or negative in value, in contrast to $I(A; B)$ which is never negative.

Furthermore, the concept of mutual information can be extended to an A^n product ensemble:

$$I(a_1; a_2; \ldots; a_n) \triangleq \log_2 \frac{[\pi p(a_i, a_j)][\pi p(a_i, a_j, a_k, a_l)] \cdots}{[\pi p(a_i)][\pi p(a_i, a_j, a_k)] \cdots} \quad (1.55)$$

where π denotes the products over all possible combinations. Furthermore, Eq. (1.55) can be written

$$I(a_1; a_2; \ldots; a_n) = I(a_1; a_2; \ldots; a_{n-1}) - I(a_1; a_2; \ldots; a_{n-1}/a_n) \quad (1.56)$$

The average mutual information is therefore

$$\begin{aligned} I(A_1; A_2; \ldots; A_n) &= \sum_{A^n} p(a_1, a_2, \ldots, a_n) I(a_1; a_2; \ldots; a_n) \\ &= \sum H(A_i) - \sum H(A_i A_j) + \sum H(A_i A_j A_k) \\ &\quad - \cdots (-1)^{n-1} H(A_1 A_2 \cdots A_n) \end{aligned} \quad (1.57)$$

where the summations are evaluated over all possible combinations.

In concluding this section, we remark that generalized mutual information may have interesting applications for communication channels with multiple inputs and outputs. We see, in the next few sections, that the definition of mutual information $I(A;B)$ eventually leads to a definition of *information channel capacity*. Finally, the information measures we have defined can be easily extended from a discrete space to a continuous space:

$$H(A) \triangleq -\int_{-\infty}^{\infty} p(a) \log_2 p(a) \, da \quad (1.58)$$

$$H(B) \triangleq -\int_{-\infty}^{\infty} p(b) \log_2 p(b) \, db \quad (1.59)$$

$$H(B/A) \triangleq -\int_{-\infty}^{\infty} \int_{-\infty}^{\infty} p(a, b) \log_2 p(b/a) \, da \, db \quad (1.60)$$

$$H(A/B) \triangleq -\int_{-\infty}^{\infty} \int_{-\infty}^{\infty} p(a, b) \log_2 p(a/b) \, da \, db \quad (1.61)$$

and

$$H(AB) \triangleq -\int_{-\infty}^{\infty} \int_{-\infty}^{\infty} p(a, b) \log_2 p(a/b) \, da \, db \quad (1.62)$$

where the p's are the probability density distributions.

1.3 COMMUNICATION CHANNELS

In the preceding sections, we discussed the measure of information and we noted that the logarithmic measure of information was the basic starting point used by Shannon in the development of information theory. We pointed out that the main objective of the Shannon information theory is efficient utilization of a communication channel. Therefore, in this section, we turn our attention to the problem of transmission of information through a prescribed communication channel with certain noise disturbances.

As noted in regard to Fig. 1.2, a communication channel can be represented by an input-output block diagram. Each of the input events a can be transformed into a corresponding output event b. This transformation of an input event to an output event may be described by a transitional (conditional) probability $p(b/a)$. Thus we see that the input-output ensemble description of the transitional probability distribution $P(B/A)$ characterizes the channel behavior. In short, the conditional probability $P(B/A)$ describes the random noise disturbances in the channel.

Communication channels are usually described according to the type of input-output ensemble and are considered *discrete* or *continuous*. If both the input and output of the channel are discrete events (discrete spaces), then the channel is called a discrete channel. But if both the input and output of the channel are represented by continuous events (continuous spaces), then the channel is called a continuous channel. However, a channel can have a discrete input and a continuous output, or vice versa. Then, accordingly, the channel is called a discrete-continuous or continuous-discrete channel.

The terminology of the concept of discrete and continuous communication channels can also be extended to spatial and temporal domains. This concept is of particular importance in an optical spatial channel, which is discussed in Chap. 3. An input-output optical channel can be described by input and output spatial domains, which can also be functions of time.

As noted in the Sec. 1.2, a communication channel can have multiple inputs and multiple outputs. Thus if the channel possesses only a single input terminal and a single output terminal, it is a *one-way channel*. However, if the channel possesses two input terminals and two output terminals, it is a *two-way channel*. In addition, one can have a channel with n input and m output terminals.

Since a communication channel is characterized by the input-output transitional probability distribution $P(B/A)$, if the transitional probability distribution remains the same for all successive input and output events, then the channel is a *memoryless channel*. However, if the transitional probability distribution changes for the preceding events, whether in the input or the output, then the channel is a *memory channel*. Thus, if the memory is finite, that is, if the transitional probability depends on a finite number of preceding events, then the channel is a *finite-memory* channel. Furthermore, if the transitional probability distribution depends on stochastic processes and the stochastic processes are assumed to be nonstationary, then the channel is a *nonstationary channel*. Similarly, if the stochastic processes the transitional probability depends on are *stationary*, then the channel is a *stationary channel*. In short, a communication channel can be fully described by the characteristics of its transitional probability distribution, for example, a discrete nonstationary memory channel.

Since a detailed discussion of various communication channels is beyond the scope of this book, we evaluate two of the simplest, yet important channels, namely, memoryless discrete channel and continuous channel.

1.4 MEMORYLESS DISCRETE CHANNELS

One of the simplest communication channels is the memoryless one-way discrete channel. Again we denote the input ensemble by A and the output ensemble by B. To characterize the behavior of the channel, we give the corresponding transitional probability distribution $P(b/a)$. We denote any one of the ith input events of A by α_i, and the corresponding output event of B by β_j. Let the input to the channel be a sequence of n arbitrary events of A:

$$\alpha^n = \alpha_1 \alpha_2 \cdots \alpha_n \qquad (1.63)$$

and the corresponding output sequence be

$$\beta^n = \beta_1 \beta_2 \cdots \beta_n \qquad (1.64)$$

where α_i and β_j are any one of the input and output events of A and B, respectively.

Since the transitional probabilities for a memoryless channel do not depend on the preceding events, the composite transitional probability is

$$P(\beta^n/\alpha^n) = P(\beta_1/\alpha_1) P(\beta_2/\alpha_2) \cdots P(\beta_n/\alpha_n) \qquad (1.65)$$

The joint probability if the output sequence β^n is

$$P(\beta^n) = \sum_{A^n} P(\alpha^n) P(\beta^n/\alpha^n) \qquad (1.66)$$

where the summation is over the A^n product space.

Introduction to Information Transmission

From the preceding sections, the average mutual information between the input and output sequences of α^n and β^n can be written

$$I(A^n; B^n) = H(B^n) - H(B^n/A^n) \tag{1.67}$$

where B^n is the output product space.

We also see that, from Eq. (1.32), the entropy of B^n can be written

$$\begin{aligned} H(B^n) &= -\sum_{B^n} P(B^n) \log_2 P(B^n) \\ &= H(B_1) + H(B_2/B_1) + H(B_3/B_2 B_1) + \cdots + H(B_n/B_{n-1} \cdots B_1) \end{aligned} \tag{1.68}$$

where

$$H(B_i/B_{i-1} \ldots B_1) \triangleq -\sum_{B^i} P(\beta_i) \log_2 P(\beta_i/\beta_{i-1} \cdots \beta_1) \tag{1.69}$$

The conditional entropy of B^n given A^n can be written

$$H(B^n/A^n) = -\sum_{A^n} \sum_{B^n} P(\alpha^n) P(\beta^n/\alpha^n) \log_2 P(\beta^n/\alpha^n) \tag{1.70}$$

From Equation (1.70), it can be shown that

$$H(B^n/A^n) = \sum_{i=1}^{n} H(B_i/A_i) \tag{1.71}$$

where

$$H(B_i/A_i) \triangleq -\sum_{A_i} \sum_{B_i} P(\alpha_i) P(\beta_i/\alpha_i) \log_2 P(\beta_i/\alpha_i) \tag{1.72}$$

$H(B_i/A_i)$ is the conditional entropy of the *i*th output event β_i given the input event α_i.

By substitution of Eqs. (1.68) and (1.71) into Eq. (1.67), we have

$$I(A^n; B^n) = \sum_{i=1}^{n} [H(B_i/B_{i-1} \cdots B_1) - H(B_i/A_i)] \tag{1.73}$$

From the definition of the average mutual information in the preceding section, we see that $I(A^n; B^n)$ measures the amount of information, on the average, provided by the *n* output events about the given *n* input events. Therefore $I(A^n; B^n)/n$ is the amount of mutual information, on the average, per event. Moreover, the channel is assumed to be memoryless, that is, $P(b/a)$ are independent of the preceding event. Thus $I(A^n; B^n)/n$ is a function of $P(\alpha^n)$ and *n*. Therefore the capacity of the channel is the maximum value of $I(A^n; B^n)/n$ for a possible probability distribution of the input sequences $P(\alpha^n)$ and length n, that is, the capacity of the channel can be defined:

$$C \triangleq \max_{P(\alpha_n), n} \frac{I(A^n; B^n)}{n} \quad \text{bits/event} \tag{1.74}$$

It is also noted that, if the input events are statistically independent (i.e., from a memoryless information source), then the channel capacity of Eq. (1.74) can be written

$$C \triangleq \max_{P(a)} I(A; B) \quad \text{bits/event} \tag{1.75}$$

That is, the channel capacity is the maximization of $I(A;B)$ over the input probability distribution of $P(a)$.

It is emphasized that evaluation of the channel capacity of Eq. (1.74) or (1.75) is by no means simple; it can be quite involved. However, we illustrate a few examples in which the maximization is particularly simple. In this case, we restrict ourselves to *discrete uniform channels*. It is generally convenient to characterize a channel by means of a *transition probability matrix*:

$$[P] = \begin{bmatrix} P(b_1/a_1) & P(b_2/b_1) & \dots & P(b_m/a_1) \\ P(b_1/a_2) & P(b_2/a_2) & \dots & P(b_m/a_2) \\ \cdot & \cdot & \dots & \cdot \\ \cdot & \cdot & \dots & \cdot \\ \cdot & \cdot & \dots & \cdot \\ P(b_1/a_n) & P(b_2/a_n) & \dots & P(b_m/a_n) \end{bmatrix} \tag{1.76}$$

With this channel matrix we can now define a uniform channel. If the rows of the transition probability matrix are permutations of identical sets of probabilities, for example, $P_1, P_2, \dots, P_m$, then the channel is said to be *uniform from input*. However, if the columns of the transition probability matrix are permutations of the same set of probabilities, then the channel is said to be *uniform from output*.

If a channel is uniform from input, then the conditional entropy $H(B/a_i)$ can be shown to be

$$H(B/a_i) = -\sum_B P(b/a_i) \log_2 P(b/a_i) = -\sum_{j=1}^{m} P_j \log_2 P_j = H(B/a) \tag{1.77}$$

which is identical for all a_i of A. The significant interpretation of Eq. (1.77) is that if the channel is uniform from input, then the transmission of any input event a of A is disturbed in the same manner by the noise in the channel. It is also noted that if the channel is uniform from output, then a given

Introduction to Information Transmission

equiprobable (i.e., uniform in probability) input ensemble of a, $[P(a) = 1/n]$ will give rise to an *equiprobable* output ensemble $[P(b) = 1/m]$.

Now if a channel is both uniform from input and output, then it is said to be a *doubly uniform channel* or just simply a *uniform channel*. In the following, we evaluate the capacity for a special type of uniform channel, namely, an *n-ary symmetric channel*.

Let the transition probability matrix of an n-ary symmetric channel be

$$[P] = \begin{bmatrix} 1-p & \frac{p}{n-1} & \frac{p}{n-1} & \cdots & \frac{p}{n-1} \\ \frac{p}{n-1} & 1-p & \frac{p}{n-1} & \cdots & \frac{p}{n-1} \\ \vdots & \vdots & \vdots & \cdots & \vdots \\ \frac{p}{n-1} & \frac{p}{n-1} & \frac{p}{n-1} & \cdots & 1-p \end{bmatrix} \tag{1.78}$$

To evaluate the channel capacity, we first evaluate the average mutual information $I(A; B)$. To do so, we calculate the conditional entropy $H(B/A)$. But from Eq. (1.77) we see that, for a channel uniform from input, the ensemble over B is independent of every a:

$$\begin{aligned} H(B/A) &= -\sum_A P(a) \sum_B P(b/a) \log_2 P(b/a) \\ &= -\sum_B P(b/a) \log_2 P(b/a) = H(B/a) \end{aligned} \tag{1.79}$$

Now we seek to maximize the average mutual information $I(A; B)$. Since $I(A; B) = H(B) - H(B/A)$, from Eq. (1.75) we see that $I(A; B)$ is maximum when $H(B)$ is maximum, and that the maximum value of $H(B)$ occurs only when every output event of b is equiprobable. However, it is noted that in general it is not true that there exists an input probability distribution of $P(a)$ such that every output event is equiprobable. But it is true for a doubly uniform channel that equiprobability of the input events produces equiprobability of the output events. Therefore the capacity of an n-ary uniform channel is

$$C = \log_2 n + \sum_B P(b/a) \log_2 P(b/a) \tag{1.80}$$

By substituting transitional probabilities of Eq. (1.78) into Eq. (1.80), we have

$$C = \log_2 n + (1-p) \log_2 (1-p) + p \log_2 \frac{p}{n-1} \tag{1.81}$$

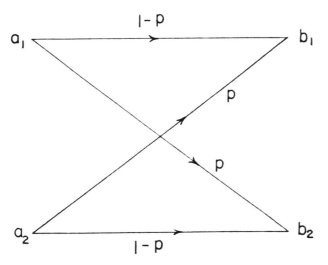

Fig. 1.4 A binary symmetric channel

which can also be written

$$C = \log_2 n - p \log_2 (n-1) - H(p) \tag{1.82}$$

where

$$H(p) \triangleq -[p \log_2 p + (1-p) \log_2(1-p)] \tag{1.83}$$

It is interesting to note that, if $n = 2$, then the n-ary channel matrix of Eq. (1.78) reduces to

$$[P] = \begin{bmatrix} 1-p & p \\ p & 1-p \end{bmatrix} \tag{1.84}$$

which is a *binary symmetric channel*, as shown in Fig. 1.4. Thus, from Eq. (1.82), the capacity of a binary symmetric channel is

$$C = 1 + p \log_2 p + (1-p) \log_2(1-p) = 1 - H(p) \tag{1.85}$$

In concluding this section, we evaluate a simple channel which is uniform from input but not from output, namely, a *binary symmetric erasure channel;*, as shown in Fig. 1.5. The corresponding transition probability matrix is

$$[P] = \begin{bmatrix} 1-p-q & p & q \\ p & 1-p-q & q \end{bmatrix} \tag{1.86}$$

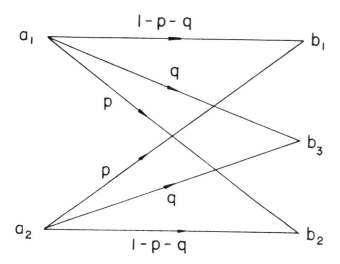

Fig. 1.5 A binary symmetric erasure channel

Since the channel is uniform from input, it follows that equiprobability of the input events produces a maximum value of the output entropy $H(B)$. It can be readily checked that the maximum value of $H(B)$ occurs when the input events are equiprobable, that is, $P(a_1) = P(a_2) = \frac{1}{2}$. Thus the output probability distribution is

$$P(b_1) = P(b_2) = \tfrac{1}{2}(1-q) \qquad P(b_3) = q \tag{1.87}$$

We see that $H(B)$ can be evaluated:

$$\begin{aligned} H(B) &= -\sum_B P(b) \log_2 P(b) \\ &= (1-q)[1 - \log_2(1-q)] - q \log_2 q \end{aligned} \tag{1.88}$$

From Eq. (1.86) we have the conditional entropy $H(B/A)$:

$$H(B/A) = -[(1-p-q)\log_2(1-p-q) + p \log_2 p + q \log_2 q] \tag{1.89}$$

Thus the channel capacity is

$$C = (1-q)[1 - \log_2(1-q)] + [(1-p-q)\log_2(1-p-q) + p \log_2 p] \tag{1.90}$$

We see that, if $p=0$, the channel capacity can be written

$$C = 1 - q \tag{1.91}$$

The capacity is equal to that of a noiseless binary channel minus the erasure probability q.

1.5 CONTINUOUS CHANNELS WITH ADDITIVE NOISE

We now consider information transmission through continuous channels. We restrict our discussion mainly to the case of additive noise.

A channel is said to be continuous if and only if the input and output ensembles are represented by continuous Euclidean spaces. For simplicity, we restrict our discussion to only the one-dimensional case, although it can be easily generalized to a higher dimension.

Again, we denote by A and B the input and output ensembles, but this time A and B are continuous random variables of a communication channel, as in Fig. 1.2. It is also noted that a continuous channel can be either *time-discrete* or *time-continuous*. We first discuss time-discrete channels and then consider time-continuous channels.

Like a discrete channel, a continuous channel is said to be memoryless if and only if its transitional probability density $p(b/a)$ remains the same for all successive pairs of input and output events. A memoryless continuous channel is said to be disturbed by an *additive noise* if and only if the transitional probability density $p(b/a)$ depends only on the difference between the output and input random variables, $b - a$:

$$p(b/a) = p(c) \tag{1.92}$$

where $c = b - a$.

Thus for additive channel noise the conditional entropy $H(B/a)$ can be shown:

$$H(B/a) = -\int_{-\infty}^{\infty} p(b/a) \log_2 p(b/a)\, db = -\int_{-\infty}^{\infty} p(c) \log_2 p(c)\, dc \tag{1.93}$$

We see that $H(B/a)$ is independent of a, which is similar to the fact of Eq. (1.79), where the channel is uniform from input. The average conditional entropy is

$$\begin{aligned} H(B/A) &= \int_{-\infty}^{\infty} p(a) H(B/a)\, da \\ &= -\int_{-\infty}^{\infty} p(c) \log_2 p(c)\, dc = H(B/a) \end{aligned} \tag{1.94}$$

It is noted that the definition of capacity for a discrete channel can also be applied to a continuous channel:

$$C \triangleq \max_{p(a)} I(A; B) \tag{1.95}$$

Thus in evaluation of the channel capacity, we evaluate first the average mutual information:

$$I(A; B) = H(B) - H(B/A) \tag{1.96}$$

where

$$H(B) \triangleq -\int_{-\infty}^{\infty} p(b) \log_2 p(b) \, db \tag{1.97}$$

and

$$H(B/A) \triangleq -\int_{-\infty}^{\infty} \int_{-\infty}^{\infty} p(a)p(b/a) \log_2 p(b/a) \, da \, db \tag{1.98}$$

We see that, from Eq. (1.94), $H(B/A)$ depends only on $P(b/a)$. Thus, if one maximizes $H(B)$, then $I(A; B)$ is maximized. But it is noted that $H(B)$ cannot be made infinitely large, since $H(B)$ is always restricted by certain physical constraints, namely, the available power. This power constraint corresponds to the mean-square fluctuation of the input signal (e.g., a mean-square current fluctuation):

$$\sigma_a^2 = \int_{-\infty}^{\infty} a^2 p(a) \, da \tag{1.99}$$

Without loss of generality, we assume that the average value of the additive noise is zero:

$$\bar{c} = \int_{-\infty}^{\infty} c p(c) \, dc = 0 \tag{1.100}$$

Then the mean-square fluctuation of the output signal can be written

$$\sigma_b^2 = \int_{-\infty}^{\infty} b^2 p(b) \, db \tag{1.101}$$

Since $b = a + c$ (i.e., signal plus noise), one can show that

$$\sigma_b^2 = \int_{-\infty}^{\infty} \int_{-\infty}^{\infty} (a+c)^2 p(a) p(c) \, da \, dc = \sigma_a^2 + \sigma_c^2 \tag{1.102}$$

where

$$\sigma_c^2 = \int_{-\infty}^{\infty} c^2 p(c) \, dc \tag{1.103}$$

From Eq. (1.102) we see that, setting an upper limit to the mean-square fluctuation of the input signal is equivalent to setting an upper limit to the mean-square fluctuation of the output signal. Thus, for a given mean-square value of σ_b^2, one can show that for the corresponding entropy, derived from $p(b)$, there exists an upper bound:

$$H(B) \leq \tfrac{1}{2} \log_2 (2\pi e \sigma_b^2) \tag{1.104}$$

where the equality holds if and only if that the probability density $p(b)$ has a Gaussian distribution, with zero mean and the variance equal to σ_b^2.

Since from the additivity property of the channel noise $H(B/A)$ depends solely on $p(c)$ [see Eq. (1.94)], we see that

$$H(B/A) \leq \tfrac{1}{2} \log_2 (2\pi e \sigma_c^2) \tag{1.105}$$

where the equality holds when $p(c)$ has a Gaussian distribution, with zero mean and the variance equal to σ_c^2.

Thus, if the additive noise in a memoryless continuous channel has a Gaussian distribution, with zero mean and the variance equal to N, where N is the average noise power, then the average mutual information satisfies the inequality

$$I(A;B) \leq \tfrac{1}{2} \log_2 (2\pi e \sigma_b^2) - \tfrac{1}{2} \log_2 (2\pi e N) \leq \tfrac{1}{2} \log_2 \frac{\sigma_b^2}{N} \tag{1.106}$$

Since the input signal and the channel noise are assumed to be statistically independent,

$$\sigma_b^2 = \sigma_a^2 + \sigma_c^2 = \sigma_a^2 + N \tag{1.107}$$

Therefore, Eq. (1.106) can also be written

$$I(A;B) \leq \tfrac{1}{2} \log_2 \frac{\sigma_a^2 + N}{N} \tag{1.108}$$

In calculating the channel capacity, we maximize the average mutual information of Eq. (1.108). We see that the equality of Eq. (1.108) holds if and only if the input signal is *also* Gaussianly distributed, with zero mean and the variance equal to S:

$$C = \tfrac{1}{2} \log_2 \left(1 + \frac{S}{N}\right) \tag{1.109}$$

It is noted that Eq. (1.109) is one of the formulas obtained by Shannon for an additive Gaussian channel.

However, it should be cautioned that, if the additive noise does *not* have a Gaussian distribution, then in general there does not exist an input probability density distribution of $p(a)$ so that the corresponding output probability density of $p(b)$ has a Gaussian distribution.

To define the *entropy power*, we let the equality hold for Eq. (1.104) and replace σ_b^2 with $\bar{\sigma}_b^2$. Thus, for a given value of $H(B)$, the entropy power of B is

$$\bar{\sigma}_b^2 \triangleq \frac{1}{2\pi e} 2^{2H(B)} \qquad (1.110)$$

We see that $\bar{\sigma}_b^2$ is equivalent to the variance of a random variable having a Gaussian distribution, with the entropy $H(B)$. Thus, from Eq. (1.104), we see that

$$\sigma_b^2 \geq \bar{\sigma}_b^2 \qquad (1.111)$$

That is, the variance of the output random variable is greater than its entropy power.

Since $b = a + c$, the sum of two statistically independent random variables a and c, from Eq. (1.111) we have the inequality

$$\bar{\sigma}_a^2 + \bar{\sigma}_c^2 \leq \bar{\sigma}_b^2 \leq \sigma_a^2 + \sigma_c^2 \qquad (1.112)$$

where $\bar{\sigma}_a^2$ and $\bar{\sigma}_c^2$ are the entropy powers of the input ensemble A and of the additive noise ensemble c. The equality of Eq. (1.112) holds if and only if the input signal and additive noise both are Gaussianly distributed, with zero means and variances equal to σ_a^2 and σ_c^2, respectively.

We now consider a memoryless continuous channel disturbed by an additive but *non-Gaussian* noise, with zero mean, variance equal to N, and entropy power $\bar{\sigma}_c^2$. If the mean-square fluctuation of the input signal cannot exceed a certain value S, then, from Eq. (1.112), one can show that the channel capacity is bounded from below and from above:

$$\tfrac{1}{2} \log_2 \left(1 + \frac{S}{\bar{\sigma}_c^2}\right) \leq C \leq \tfrac{1}{2} \log_2 \frac{N + S}{\bar{\sigma}_c^2} \qquad (1.113)$$

The equality holds if and only if the additive channel noise is Gaussianly distributed, with zero mean and the variance equal to N, that is $\bar{\sigma}_c^2 = N$.

Furthermore, in view of Eq. (1.105), we see that the noise entropy is maximum (for a given value of σ_c^2) if the random noise c is Gaussianly distributed, with zero mean. Thus the noise disturbance is expected to be more severe for additive Gaussian noise. Therefore, from Eq. (1.113) and the fact that $\sigma_c^2 \geq \bar{\sigma}_c^2$, the capacity of a memoryless continuous channel disturbed by additive non-Gaussian noise with zero mean and the variance equal to N, where the mean-square fluctuation of the input signal does not exceed a given value of S, is

$$C \geq \tfrac{1}{2} \log_2 \left(1 + \frac{S}{N}\right) \tag{1.114}$$

which is larger than the capacity of an additive Gaussian channel [Eq. (1.109)]. We also see that the equality holds for Eq. (1.114) if and only if the additive noise has a Gaussian distribution, with zero mean and the variance equal to N.

We can now evaluate the most well-known channel in information theory, namely, a memoryless, time-continuous, band-limited, continuous channel. The channel is assumed to be disturbed by an additive white Gaussian noise, and a band-limited time-continuous signal, with an average power not to exceed a given value S, is applied at the input end of the channel.

It is noted that, if a random process is said to be a stationary Gaussian process, then the corresponding joint probability density distribution, assumed by the time functions at any finite time interval, is independent of the time origin selected, and it has a Gaussian distribution. If a stationary Gaussian process is said to be white, then the power spectral density must be uniform (constant) over the entire range of the frequency variable. An example, which we encounter later, is thermal noise which is commonly regarded as having a *stationary* Gaussian distribution and frequently is assumed to be white.

To evaluate the capacity of an additive stationary Gaussian channel, we first illustrate a basic property of white Gaussian noise. Let $c(t)$ be a white Gaussian noise; then by the Karhunen-Loéve expansion theorem, [14,15], $c(t)$ can be written over a time interval $-T/2 \leq t < T/2$:

$$c(t) = \sum_{i=-\infty}^{\infty} c_i \phi_i(t) \tag{1.115}$$

where the $\phi_i(t)$'s are *orthonormal functions*, such that

$$\int_{-T/2}^{T/2} \phi_i(t)\phi_j(t)\, dt = \begin{cases} 1 & i=j \\ 0 & i \neq j \end{cases} \tag{1.116}$$

and c_i are real coefficients commonly known as *orthogonal expansion coefficients*. Furthermore the c_i's are statistically independent, and the individual probability densities have a stationary Gaussian distribution, with zero mean and the variances equal to $N_0/2T$, where N_0 is the corresponding power spectral density.

Now we consider an input time function $a(t)$, applied to the communication channel, where the frequency spectrum is limited by some bandwidth Δv of the channel. Since the channel noise is assumed to be additive white Gaussian noise, the output response of the channel is

$$b(t) = a(t) + c(t) \tag{1.117}$$

Such a channel is known as a band-limited channel with additive white Gaussian noise.

Again by the Karhunen-Loéve expansion theorem, the input and output time functions can be expanded:

$$a(t) = \sum_{i=-\infty}^{\infty} a_i \phi_i(t) \tag{1.118}$$

and

$$b(t) = \sum_{i=-\infty}^{\infty} b_i \phi_i(t) \tag{1.119}$$

Thus we see that

$$b_i = a_i + c_i \tag{1.120}$$

Since the input function $a(t)$ is band-limited by Δv, only $2T\Delta v$ coefficients a_i, $i = 1, 2, \ldots, 2T\Delta v$ within the passband are considered (we discuss this in Chap. 2). In other words, the input signal ensemble can be represented by a $2T\Delta v$-order product ensemble over a, that is, $A^{2T\Delta v}$. Similarly, the above statement is also true for the output ensemble over b, that is, $B^{2T\Delta v}$. Thus the average amount of information between the input and output ensembles is

$$I(A^{2T\Delta v}; B^{2T\Delta v}) = H(B^{2T\Delta v}) - H(B^{2T\Delta v}/A^{2T\Delta v}) \tag{1.121}$$

It is also clear that a, b, and c each form a $2T\Delta v$-dimensional *vector space*. For convenience, we denote by **a**, **b**, and **c** the respective vectors in the vector space. Thus we see that

$$\mathbf{b} = \mathbf{a} + \mathbf{c} \tag{1.122}$$

If we let $p(\mathbf{a})$ and $p(\mathbf{c})$ be the probability density distribution of **a** and **c**, respectively, then the transitional probability density of $p(\mathbf{b}/\mathbf{a})$ is

$$p(\mathbf{b}/\mathbf{a}) = p(\mathbf{b} - \mathbf{a}) = p(\mathbf{c}) \tag{1.123}$$

where **a** and **c** are statistically independent. For simplicity, we let $X \triangleq A^{2T\Delta v}$

be the vector space (the product space) of **a**. The probability density distribution of **b** can be determined:

$$p(\mathbf{b}) = \int_X p(\mathbf{a}) p(\mathbf{c}) \, dX \tag{1.124}$$

where the integral is over the entire vector space X.

Similarly, $Y \triangleq B^{2T\Delta v}$ and $Z \triangleq C^{2T\Delta v}$ represent the vector space of **b** and **c**, respectively. The average mutual information of Eq. (1.121) can therefore be written

$$I(X; Y) = H(Y) - H(Z) \tag{1.125}$$

where

$$H(Y) = -\int_Y p(\mathbf{b}) \log_2 p(\mathbf{b}) \, dY \tag{1.126}$$

and

$$H(Z) = H(Y/X) = -\int_Z p(\mathbf{c}) \log_2 p(\mathbf{c}) \, dZ \tag{1.127}$$

In view of Eq. (1.96), we see that the problem of a time-continuous channel has been reduced to a form very similar to that of a time-discrete channel with additive Gaussian noise. Thus the channel capacity can be defined, as before, as

$$C \triangleq \max_{T, p(a)} \frac{I(X; Y)}{T} \quad \text{bits/time} \tag{1.128}$$

We can now maximize the average mutual information of $I(X; Y)$ under the constraint that the mean-square fluctuation of the input signal ensemble cannot exceed a specified value S:

$$\int_X |\mathbf{a}|^2 p(\mathbf{a}) \, dX \leq S \tag{1.129}$$

Since each of the vectors **a**, **b**, and **c** can be represented by $2T \Delta v$ continuous variables, and each c_i is statistically independent, with Gaussian distribution, zero mean, and the variance equal to $N_0/2T$, we quickly see that

$$I(X; Y) = I(A^{2T\Delta v}; B^{2T\Delta v}) = \sum_{i=1}^{2T\Delta v} I(A_i; B_i) \tag{1.130}$$

Thus, from Eq. (1.105),

$$H(Z) = 2T \, \Delta v \, H(C_i) \tag{1.131}$$

where
$$H(C_i) = \tfrac{1}{2} \log_2 (2\pi e \sigma_{c_i}^2) \tag{1.132}$$

the entropy of any one of the c_i variables. If we let $N = \sigma_{c_i}^2 = N_0 \Delta v$, then we have

$$H(Z) = T \Delta v \log_2 \left(\frac{\pi e N_0}{T} \right) \tag{1.133}$$

In view of Eq. (1.104), we see that

$$H(B_i) \leq \tfrac{1}{2} \log_2 (2\pi e \sigma_{b_i}^2) \tag{1.134}$$

with the equality holding if and only if b_i is Gaussianly distributed, with zero mean and the variance equal to $\sigma_{b_i}^2$. Since $\mathbf{b} = \mathbf{a} + \mathbf{c}$, we see that for $p(\mathbf{b})$ to have a Gaussian distribution $p(\mathbf{a})$ also must have a Guassian distribution, with zero mean. Thus the average mutual information of Eq. (1.130) can be written

$$\begin{aligned} I(X;Y) &= \sum_{i=1}^{2T\Delta v} H(B_i) - H(Z) \\ &= \tfrac{1}{2} \log_2 \left[\prod_{i=1}^{2T\Delta v} (2\pi e \sigma_{b_i}^2) \right] - T \Delta v \log_2 \left(\frac{\pi e N_0}{T} \right) \end{aligned} \tag{1.135}$$

where Π denotes the product ensemble. Since a_i and c_i are statistically independent,

$$\sigma_{b_i}^2 = \sigma_{a_i}^2 + \sigma_{c_i}^2 = \sigma_{a_i}^2 + \frac{N_0}{2T} \tag{1.136}$$

In view of Eq. (1.129), we see that

$$\sum_{i=1}^{2T\Delta v} \sigma_{b_i}^2 = \sum_{i=1}^{2T\Delta v} \sigma_{a_i}^2 + N_0 \Delta v \leq S + N \tag{1.137}$$

where $N = N_0 \Delta v$. The equality holds for Eq. (1.137) when the input probability density distribution $p(\mathbf{a})$ has a Gaussian distribution, with zero mean and the variance equal to S. Furthermore, from Eq. (1.137), we can write

$$\prod_{i=1}^{2T\Delta v} \sigma_{b_i}^2 \leq \left(\frac{S+N}{2T \Delta v} \right)^{2T\Delta v} \tag{1.138}$$

where the equality holds if and only if the $\sigma_{b_i}^2$'s are all equal and $p(\mathbf{a})$ has a Gaussian distribution, with zero mean and the variance equal to S.

Therefore, in the maximization of Eq. (1.135), the corresponding channel capacity is

$$C = \max_{T, p(a)} \frac{I(X;Y)}{T} = \Delta v \log_2\left(1 + \frac{S}{N}\right) \quad \text{bits/sec} \qquad (1.139)$$

where S and N are the average signal and average noise power, respectively. This is one of the most popular results derived, by Shannon [3] and independently by Wiener [5], for the memoryless additive Gaussian channel. Because of its conceptual and mathematical simplicity, this equation has been frequently used in practice, but it has been occasionally misused. It is noted that this channel capacity is derived under the additive white Gaussian noise regime, and the average input signal power cannot exceed a specified value of S. As noted, we can obtain the capacity of the channel if and only if the input signal is also Gaussianly distributed, with zero mean and the variance equal to S

Since the average noise power over the specified bandwidth of the channel is $N_0 \Delta v$, we see that, for a fixed value of S/N, as the channel bandwidth increases to infinitely large, the capacity of the channel approaches a definite value:

$$C(\infty) = \lim_{\Delta v \to \infty} C(\Delta v) = \frac{S}{N_0} \log_2 e \qquad (1.140)$$

This equation possesses an important physical significance: The measurement or observation of any physical quantity is practically always limited by *thermal noise agitation*. This thermal agitation can usually be considered white (at least within the bandwidth of interest) additive Gaussian noise. The noise power spectral density N_0 can be related to the thermal temperature T:

$$N_0 = kT \qquad (1.141)$$

where k is Boltzmann's constant, and T is in Kelvins. Moreover, from Eq. (1.140), it follows that the signal energy transmitted through a physical communication channel must be at least kT per nat of information the signal is able to provide. In other words, it takes at least kT energy for a nat of information to be properly transmitted.

In concluding this section, we plot the capacity of the additive Gaussian channel as a function of bandwidth, for a fixed value of S/N_0, as shown in Fig. 1.6. We see that, for small values of channel bandwidth Δv, the capacity increases very rapidly with Δv, but that it asymptotically approaches $C(\infty)$ of Eq. (1.140) when Δv becomes infinitely large.

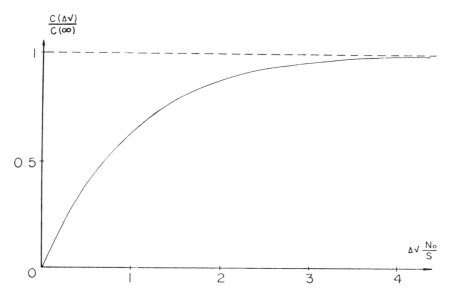

Fig. 1.6 The capacity of an additive white Gaussian channel as a function of bandwidth Δv.

1.6 SUMMARY AND REMARKS

We devoted the earlier part of this chapter to the definition of information measures. With these basic definitions it is possible to study the important properties of transmission of information. More precisely, the measure of information was defined as the logarithm of the ratio of the a posteriori probability to the priori probability of an event. We showed that such an information measure possesses a symmetric property such that the measure of mutual information can be defined.

Moreover, self-information can also be defined as the logarithm of the *a priori* probability of an event. Thus the measure of self-information of an event is the amount of information required to specify that event *uniquely*. In other words, it is the amount of information the event can provide.

Furthermore, with these primary information measures, the ensemble average of information measures can be determined, namely, the average self-information and the average mutual information. We have shown that the average mutual information can never be negative, in contrast to mutual information about a particular event, which can be either positive or negative. However, the greatest significance of the average mutual information is that it leads us to the basic definition and evaluation of channel capacity.

It is also noted that the information measures we have defined were originally applied on discrete space. However, the definition can be easily extended to continuous space.

With the entropy equations, we have evaluated simple communication channels. In the evaluation of channel capacity, one can maximize the corresponding average mutual information of the channel by searching for an optimum input probability distribution. In the maximization of the mutual information it is required that the input events be statistically independent. As a result, the output events are also statistically independent.

After evaluation of simple memoryless discrete channels, we turned our attention to memoryless continuous channels with an additive noise regime. However, the capacities of continuous channels were evaluated only in some special, although important, cases, namely, for additive Gaussian noise. The case for additive non-Gaussian noise was also discussed, and we showed that the capacity has a lower and an upper bound.

We also evaluated the capacity of a memoryless time-continuous channel under the additive noise regime and arrived at an important and interesting formula for the channel capacity. This formula was first derived independently by Wiener and by Shannon. Because of its conceptual and mathematical simplicity, this formula has been frequently used by communication engineers. We should remember that this formula was derived under the assumption of an additive white Gaussian noise regime and to approach the capacity only through a Gaussian signal.

It is emphasized that the evaluation of a communication channel with memory is very laborious and extremely difficult. To our knowledge no satisfactory result has been achieved for a general memory channel. However researchers in this area are still very active, and certainly important results for memory channels will eventually be obtained.

In concluding this brief summary, we state a very important fundamental theorem without proof: If a communication channel (noisy or non-noisy) possesses a channel capacity of C bits per second and is assumed to accept an input signal from an information source at a rate of H bits per second, where $H \leq C$, then by properly encoding the signal it is possible to transmit it through the channel with as little error as we please. However, if the rate of information is greater than the channel capacity, $H > C$, then it is impossible to code the source signal so that it can be transmitted through the channel with as small an error as we please.

Finally, it should be clear that the purpose of this chapter is not to cover the entire domain of information transmission but rather to present a basic introduction. The topics selected for this text were those of general interest. Therefore, readers interested in a detailed discussion on information transmission can refer to the original papers by Shannon [1–3] and the excellent texts by Fano [4] and by Gallager [16].

REFERENCES

1. C. E. Shannon, "A Mathematical Theory of Communication," *Bell Syst. Tech. J.*, vol. 27, 379–423, 623–656 (1948).
2. C. E. Shannon, "Communication in the Presence of Noise," *Proc. IRE*, vol. 37, 10 (1949).
3. C. E. Shannon and W. Weaver, *The Mathematical Theory of Communication*, University of Illinois Press, Urbana, 1949.
4. R. M. Fano, *Transmission of Information*, MIT Press, Cambridge, Mass., 1961.
5. N. Wiener, *Cybernetics*, MIT Press, Cambridge, Mass., 1948.
6. N. Wiener, *Extrapolation, Interpolation, and Smoothing of Stationary Time Series*, MIT Press, Cambridge, Mass., 1949.
7. R. V. L. Hartley, "Transmission of Information," *Bell Syst. Tech. J.*, vol. 7, 535 (1928).
8. L. Brillouin, "The Negentropy Principle of Information." *J. Appl. Phys.*, vol. 24, 1152 (1953).
9. L. Brillouin, *Science and Information Theory*, Academic, New York, 1956.
10. L. Brillouin, *Scientific Uncertainty and Information*, Academic, New York, 1964.
11. D. Gabor, "Light and Information," in E. Wolf, ed., *Progress in Optics*, vol. I, North-Holland, Amsterdam, 1961.
12. D. Gabor, "Informationstheorie in der Optik," *Optik*, vol. 39, 86 (1973).
13. W. J. McGill, "Multivariate Information Transmission," *IRE Trans. Inf. Theory*, vol. 4, 93 (1954).
14. W. Davenport and W. Root, *Random Signals and Noise*, McGraw-Hill, New York, 1958.
15. M. Loéve, *Probability Theory*, 3rd ed., Van Nostrand, Princeton, N.J., 1963.
16. R. G. Gallager, *Information Theory and Reliable Communication*, John Wiley, New York, 1968.

2
Diffraction and Signal Analysis

Since the first word in the title of this chapter is diffraction, the first question is, naturally, What is diffraction? The answer to this question may be simplified as follows. When light passes the edge of an obstacle, its course deviates from that of a straight line. This deviation effect is known as *diffraction*.

Historically, it was the observation of diffraction that ultimately led to the acceptance of the wave theory of light. Thus it is possible to explain diffraction on the basis of wave theory. For example, the angle of deviation from straight-line propagation, it can be shown, is proportional to the wavelength of the light source. However, the wavelengths of visible light are extremely small, and thus the diffraction angles are also very small. We see that in geometric optics light propagation is assumed to occur in a straight line. Therefore we note that, in diffraction theory, straight-line propagation occurs only at the limit where the wavelength is zero.

It is emphasized that the theory of diffraction can be treated on the basis of Huygens's principle, in the form of Kirchhoff's integral. Although rigorous analysis in diffraction should be obtained from Maxwell's equations, the scalar theory of Huygens provides very good predicted results, as in the effect of the actual diffraction. Thus scalar theory provides a good qualitative analysis of diffraction phenomenon. However, the greatest significance of the scalar theory approach is that the analysis is much less difficult as compared with analysis by means of Maxwell's equations.

In this chapter, we provide a brief discussion of diffraction and of signal analysis. Our discussion is primarily an introductory one. Therefore, interested readers are referred to the books by Born and Wolf [1] and by Yu [2] for diffraction theory, and to the book by Davenport and Root [3] for signal analysis.

2.1 INTRODUCTION TO DIFFRACTION

It is customary to divide diffraction into two cases, depending on the distance of the light source and the viewing point from the diffracting screen, and these cases have been named after two early investigators of diffraction. If the source and viewing point are so far from the diffracting screen that lines drawn from the source of viewing point to all points of the apertures do not differ in length by more than a small fraction of a wavelength, the phenomenon is called *Fraunhofer diffraction*. If these conditions do not hold, it is called *Fresnel diffraction*. The boundary between these two cases is somewhat arbitrary and depends on the accuracy desired in the results. In most cases it is sufficient to use Fraunhofer methods if the difference in distances does not exceed one-twentieth of a wavelength. Of course, Fraunhofer diffraction can be achieved without the source being at a great physical distance if a collimating lens is used to make the rays of light from the source nearly parallel.

Figure 2.1 illustrates the above considerations. A source of monochromatic light is located at S, and a viewing point at P, and between them is an opaque screen having a finite number of apertures. Let a circle C be drawn on the plane of the screen, and let it be as small as possible while still enclosing all the apertures. Let C be the base of cones with S and P as vertices. Also draw spherical surfaces Σ_1 and Σ_2, having S and P as centers, and as radii r_1 and r_2, the shortest distances from S and P to the base C. If the longest distances from C to Σ_1 and to Σ_2 are not more

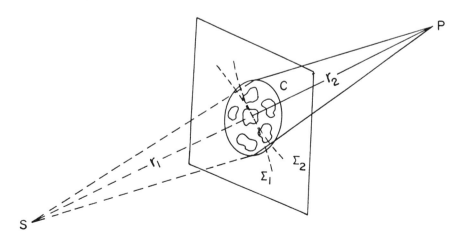

Fig. 2.1 Geometry for defining Fraunhofer and Fresnel diffraction.

Diffraction and Signal Analysis   37

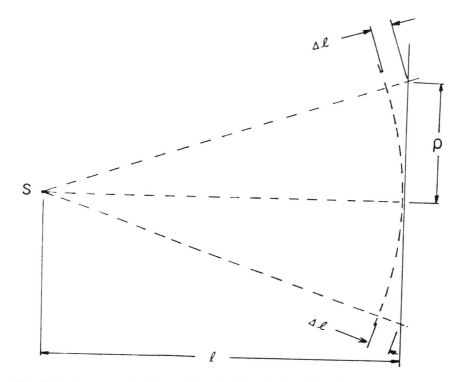

Fig. 2.2 Geometry for determining the Fraunhofer diffraction.

than one-twentieth of the wavelength of the light used, then the diffraction is Fraunhofer, and the light falling on the observing screen at P forms a *Fraunhofer diffraction pattern*.

However, if, by reason of the large size of C or the shortness of the distances to S and P, the distances between C and Σ_1 or Σ_2 are greater than one-twentieth of the wavelength, then we have Fresnel diffraction and a *Fresnel diffraction pattern* at P.

The radius of the circle C in Fig. 2.2 is denoted by ρ, the shortest distance to S from the screen is l, and the greatest separation of the sphere and the screen is Δl. From the definition of Fraunhofer diffraction, Δl must be a small fraction of a wavelength. However, ρ may be many wavelengths long (as it can also be in the Fresnel case). In the right triangle in Fig. 2.2, we have

$$(l + \Delta l)^2 = l^2 + \rho^2 \qquad (2.1)$$

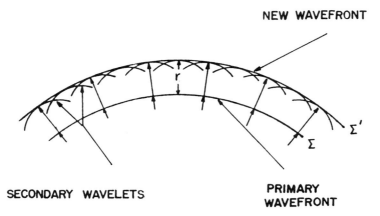

Fig. 2.3 Huygens' principle.

and because of the small size of $(\Delta l)^2$ in comparison with the other quantities, we make the approximation

$$l \simeq \frac{\rho^2}{2\,\Delta l} \tag{2.2}$$

As an example, suppose that ρ is 1 cm, and that the source has the longest wavelength in the visible red, about 8×10^{-5} cm. Let Δl be one-twentieth of this, or 0.4×10^{-5} cm. The distance l is then approximately 1.25 km. If violet light of half the wavelength is used, l will be 2.5 km. The requirements for Fraunhofer diffraction can thus be rather stringent.

We now turn our attention to one of the most significant principles in the theory of diffraction, namely, Huygens' principle. By means of Huygens' principle it is possible to obtain by graphical methods the shape of a wave front at any instant if the wave front at an earlier instant is known. The principle may be stated as follows. Every point of a wave front may be considered the source of a small secondary wavelet which spreads in all directions from the point at the wave propagation velocity. A new wave front is found by constructing a surface tangent to all the second wavelets. If the velocity of propagation is not constant for all parts of the wave front, then each wavelet must be given an appropriate velocity.

An illustration of the use of Huygens' principle is given in Fig. 2.3. The known wave front is shown by the arc Σ, and the directions of propagation are indicated by small arrows. To determine the wave front after an interval Δt, with a wave velocity v, simply construct a series of spheres of radius $r = v\Delta t$ from each point of the original front Σ. These spheres represent the secondary wavelets. The envelope enclosing the surfaces of the spheres represents the new wave front. This is the surface marked Σ' in the figure. In this example the magnitude of the wave velocity is considered the same at all points.

It is noted that Huygens' principle predicts the existence of a backward wave which, however, is never observed. An explanation for this discrepancy can be found by examining the interference of the secondary wavelets throughout the space surrounding Σ. The main use of Huygens' principle is in predicting diffraction patterns, which we see in Sec. 2.2. When first stated the principle was a useful method for finding the shape of a new wave front; little physical significance was attached to the secondary wavelets at that time. Later, as the wave nature of light came to be more fully understood, Huygens' principle took on deeper significance.

2.2 FRESNEL-KIRCHHOFF THEORY

In Sec. 2.1 we discussed the basic concept of diffraction. We defined far-field Fraunhofer diffraction and near-field Fresnel diffraction. The basic concept of Huygens' principle was also illustrated.

In this section we derive one of the most important theories in diffraction, namely, the *Fresnel-Kirchhoff theory*. We show that the Fresnel-Kirchhoff theory essentially coincides with Huygens' principle.

A rigorous approach to the Fresnel-Kirchhoff theory should start with the application of *Green's theorem to the scalar wave theory*. However, such a derivation is in general more mathematically involved, so we derive the Fresnel-Kirchhoff theory from a *system theory* viewpoint. In beginning the derivation, we recall Huygens' principle from Sec. 2.1. Thus, according to Huygens' principle, the amplitude observed at a point P' of a given coordinate system $\sigma(\alpha, \beta, \gamma)$, due to a light source located in a given coordinate system $\rho(x, y, z)$, as shown in Fig. 2.4, can be calculated by assuming that each point of the light source is an infinitesimal spherical radiator. Thus the complex light amplitude $u(r)$ contributed by a point P in the ρ coordinate system can be considered that from an unpolarized monochromatic point source, such that

$$u(r) = \frac{1}{r} \exp\left[i(kr - \omega t)\right] \tag{2.3}$$

where k and ω are the wave number, and angular frequency, respectively, of the point source, and r is the distance between the point source and the point of observation:

$$r = [(l + \gamma - z)^2 + (\alpha - x)^2 + (\beta - y)^2]^{1/2} \tag{2.4}$$

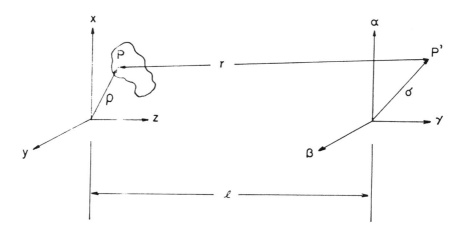

Fig. 2.4 Fresnel-Kirchhoff theory.

If the separation l of the two coordinate systems is assumed to be large compared to the magnitude of ρ and σ, then r can be approximated by l in the denominator of Eq. (2.3), and by

$$r = \left[l' + \frac{(\alpha - x)^2}{2l'} + \frac{(\beta - y)^2}{2l'}\right] \qquad (2.5)$$

in the exponent of Eq. (2.3). Since $l' = l + \gamma - z$, Eq. (2.3) can be written

$$u(\sigma - \rho) \simeq \frac{1}{l}\exp\left\{ik\left[l' + \frac{(\alpha - x)^2}{2l'} + \frac{(\beta - y)^2}{2l'}\right]\right\} \qquad (2.6)$$

where the time-dependent exponent has been omitted for convenience.

Furthermore, if the point of radiation and the point of observation are interchanged in the two coordinate systems, then the complex light amplitude observed at $\rho(x, y, z)$ is

$$u(\rho - \sigma) = \frac{1}{l}\exp\left\{ik\left[l' + \frac{(\alpha - x)^2}{2l'} + \frac{(\beta + y)^2}{2l'}\right]\right\} \qquad (2.6)$$

where $l'' = l + z - \gamma$.

It is clear that Eqs. (2.6) and (2.7) represent the free-space radiation from a monochromatic point source. They are also called free-space *impulse responses*.

Diffraction and Signal Analysis

Therefore, the complex amplitude produced at the σ coordinate system by a monochromatic radiating surface located in the ρ coordinate system can be written

$$U(\sigma) = \int\int_\Sigma T(\rho)u(\sigma - \rho)\, d\Sigma \tag{2.8}$$

where $T(\rho)$ is the complex light field of the monochromatic radiating surface, Σ denotes the surface integral, and $d\Sigma$ is the incremental surface element. It is noted that Eq. (2.8) represents the Fresnel-Kirchhoff theory.

Furthermore, by replacing $u(\sigma - \rho)$ by $u(r)$, from Eq. (2.8) we see that

$$T(\rho)u(\sigma - \rho)\, d\Sigma = T(\rho)u(r)\, d\Sigma = \frac{C\, d\Sigma}{r} e^{i(kr-\omega t)} \tag{2.9}$$

where $C = T(\rho)$. Equation (2.9) is essentially the secondary-point radiation proposed by Huygens. Thus Eq. (2.8) also represents Huygens' principle in the form of Kirchhoff's integral.

2.3 LINEAR SYSTEMS AND FOURIER ANALYSIS

The concept of a linear system is very important in the analysis of optical communication and information processing systems for at least two major reasons:

1. A great number of applications in optical communication and processing systems are assumed to be linear, at least within specified ranges.
2. An exact solution in the analysis of linear information transmission can be obtained by standard techniques.

Except for a very few special cases, there is no general procedure for analyzing nonlinear problems. Of course, there are practical ways of solving nonlinear problems which may involve graphical or experimental approaches. Approximations are often necessary in solving nonlinear problems, and each situation may require special handling techniques. In practice, fortunately, a great number of optical communication and processing problems are linear, and these are generally solvable. However, we emphasize that, in practice, no physical system is strictly linear unless certain restrictions are imposed.

It is a common practice to study the behavior of a physical system by the input excitation and output response. Both excitation and response may be some physically measurable quantity, depending on the nature

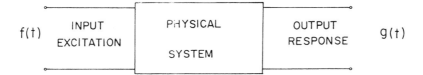

Fig. 2.5 An input-output physical system.

of the system. Suppose an input excitation of $f_1(t)$ produces an output response of $g_1(t)$, and that a second excitation $f_2(t)$ produces a second response $g_2(t)$ of a physical system, as shown in Fig. 2.5. Symbolically, we write

$$f_1(t) \to g_1(t) \tag{2.10}$$

and

$$f_2(t) \to g_2(t) \tag{2.11}$$

Then for a linear physical system we have

$$f_1(t) + f_2(t) \to g_1(t) + g_2(t) \tag{2.12}$$

Equation (2.12), in conjunction with Eqs. (2.10) and (2.11), represents the *additivity* property of the linear system. Thus a necessary condition for a system to be linear is that the *principle of superposition* hold. The principle of superposition implies that the presence of one excitation does not affect the response due to other excitations.

Now, if the input excitation of the same physical system is $Cf_1(t)$, where C is an arbitrary constant, then the output response is $Cg_1(t)$:

$$Cf_1(t) \to Cg_1(t) \tag{2.13}$$

Equation (2.13) represents the *homogeneity* characteristic of the linear system. Thus a property of a linear system is preserving the magnitude scale factor. Therefore it may be concluded that a physical system is *linear* if and only if Eqs. (2.12) and (2.13) are satisfied. In other words, if a system possesses the additivity and homogeneity properties, then it is a *linear system*.

There is, however, another important physical aspect that characterizes a linear system with constant parameters. If the input excitation f(t) applied to such a system is an alternating function of t with frequency v and if the output response $g(t)$ appears to be alternating with the same frequency v, then the system can be said to have *time invariance*.

Diffraction and Signal Analysis

In other words, a time-invariant system does not generate new frequencies. Time invariance implies that if

$$f(t) \to g(t)$$

then

$$f(t - t_0) \to g(t - t_0) \tag{2.14}$$

where t_0 is an arbitrary time delay. Therefore, if a linear system possesses the time-invariance property, then it is a *linear time-invariant system*.

We now turn to Fourier analysis. Fourier transforms are particularly important in the analysis of optical information processing and communication. We consider first a class of real functions $f(t)$ which satisfy the following sufficient conditions:

1. $f(t)$ must be sectionally continuous in every finite region over the t domain, except for a finite number of discontinuities.
2. $f(t)$ must be absolutely integrable over the entire t domain:

$$\int_{-\infty}^{\infty} |f(t)| dt < \infty \tag{2.15}$$

These functions can be represented by the equation

$$f(t) = \int_{-\infty}^{\infty} F(v) \exp(i2\pi v t) \, dv \tag{2.16}$$

where

$$F(v) = \int_{-\infty}^{\infty} f(t) \exp(-i2\pi v t) \, dt \tag{2.17}$$

and v is the frequency variable.

Equations (2.16) and (2.17) are known as a *Fourier transform pair*. Equation (2.17) is often called the *Fourier transform*, and Eq. (2.16) is known as the *inverse Fourier transform*. Symbolically, Eqs. (21.6) and (2.17) can be written, respectively, as

$$f(t) = \mathcal{F}^{-1}[F(v)] \tag{2.18}$$

and

$$F(v) = \mathcal{F}[f(t)] \tag{2.19}$$

where $\mathcal{F}^{-1}$ and $\mathcal{F}$ denote the inverse and direct Fourier transformations, respectively. It is noted that $F(v)$ is generally a complex function:

$$F(v) = |F(v)| \exp[i\phi(v)] \tag{2.20}$$

where $|F(v)|$ and $\phi(v)$ are referred to as the *amplitude spectrum* and the *phase spectrum*, respectively, and $F(v)$ is also known as the *Fourier spectrum* or *frequency spectrum*. However, if $f(t)$ is a periodic function such that (1) $f(t)=f(t+T)$, where T is the period, and (2) $f(t)$ is sectionally continuous over the period T, then $f(t)$ can be expanded into a *Fourier series*:

$$f(t) = \sum_{n=-\infty}^{\infty} C_n \exp(i2\pi n v_0 t) \tag{2.21}$$

where

$$v_0 = \frac{1}{T}$$

and

$$C_n = \frac{1}{T} \int_0^T f(t) \exp(-i2\pi n v_0 t)\, dt \tag{2.22}$$

C_n is known as the *complex Fourier coefficients*, and $|C_n|$ is known as the *amplitude spectrum*. It should be noted that, if $f(t)$ is a real function, then

$$C_{-n} = C_n^* \tag{2.23}$$

where the superscripts asterisk denotes the *complex conjugate*.

In concluding this section, we refer the reader to the excellent texts by Papoulis [4] and by Bracewell [5] for a detailed treatment of Fourier analysis

2.4 FINITE BANDWIDTH ANALYSIS

Strictly speaking, optical communication channels, optical information processors, and physical systems are restricted to a finite bandwidth. For example, a physical system can be a *low-pass*, *bandpass*, or *discrete bandpass* system. But a strictly high-pass system can never exist in practice, since every physical system has an upper frequency limit. It may be pointed out that the term *frequency* can be applied to temporal and to spatial coordinates [2], as we briefly note at the end of this section. A low-pass system is defined as a system that possesses a nonzero transfer characteristic from zero frequency to a definite limit of maximum frequency v_m. Thus the bandwidth Δv of a low-pass system is equal to the maximum frequency limit:

$$\Delta v = v_m \tag{2.24}$$

However, if the system possesses a nonzero transfer characteristic

Diffraction and Signal Analysis

from a lower frequency limit v_1 to a higher frequency limit v_2, then it is a band-pass system. Thus the corresponding bandwidth is

$$\Delta v = v_2 - v_1 \tag{2.25}$$

Similarly, one can generalize the band-pass system to a finite discrete band-pass system. Thus the bandwidths must be the corresponding passbands.

Since, as is well known, the analysis of a band-pass system can be easily reduced to the case of an equivalent low-pass system [6], we restrict our discussion to only the low-pass problem.

Before beginning a detailed discussion of the analysis, one may ask a very fundamental question: Given a low-pass communication system of bandwidth $\Delta v = v_m$, what sort of output response can be expected? In other words, from the frequency domain standpoint, what would happen if the Fourier spectrum of the input signal were extended beyond the passband?

To answer this basic question, we present a very simple but important example. For simplicity, we let the low-pass system, an ideal filter as shown in Fig. 2.6, have the corresponding transfer function:

$$H(v) = \begin{cases} 1 & |v| \leq v_m \\ 0 & |v| > v_m \end{cases} \tag{2.26}$$

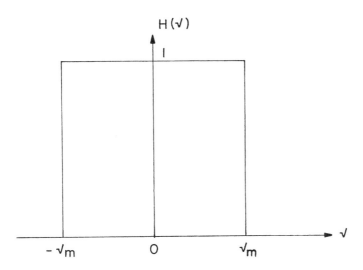

Fig. 2.6 An ideal low-pass filter.

If the input signal to this low-pass system has a finite duration of Δt, then we see that to have good output reproduction of the input signal it is required that the system bandwidth Δv be greater than or at least equal to $1/\Delta t$:

$$\Delta v \geq \frac{1}{\Delta t} \tag{2.27}$$

where $1/\Delta t$ is known as the input signal bandwidth. Thus we have the basic relationship

$$\Delta t \, \Delta v \geq 1 \tag{2.28}$$

where $\Delta v = {}^2 v_m$. We see that Eq. (2.28), the duration bandwidth product, has a definite lower bound. The basic significance is that, if the Fourier spectrum of the input signal is more or less concentrated in the passband of the system, that is, if $|v| \leq v_m$, then the output response will quite faithfully reproduce the input signal. However, if the Fourier spectrum of the input signal spreads beyond the passband of the system, then the output response will be severely distorted, that is, it will fail to reproduce the input signal. It is noted that Eq. (2.28) possesses an intimate relationship to Heisenberg's uncertainty principle in quantum mechanics, as we elaborate shortly.

We now give a few more examples in which we show that the inequality of Eq. (2.28) is indeed relevant in information transmission. Let us consider a short symmetric pulse with a duration of Δt, which has its maximum value at $t=0$:

$$f(t) = \begin{cases} f(t) & |t| \leq \dfrac{\Delta t}{2} \\ 0 & |t| > \dfrac{\Delta t}{2} \end{cases} \tag{2.29}$$

$$f(-t) = f(t) \tag{2.30}$$

and

$$f(0) \geq f(t) \tag{2.31}$$

Since the symmetric pulse $f(t)$ is Fourier-transformable, we have the corresponding Fourier spectrum of $f(t)$:

$$F(v) = \mathcal{F}[f(t)] \tag{2.32}$$

where $\mathcal{F}$ denotes the direct Fourier transformation.

We define a *nominal duration* of Δt equivalent to the duration of a rectangular pulse of amplitude $f(0)$:

$$\Delta t f(0) \triangleq \int_{-\infty}^{\infty} f(t) \, dt \tag{2.33}$$

Similarly, one can also define a *nominal bandwidth* of Δv:

$$\Delta v F(0) \triangleq \int_{-\infty}^{\infty} F(v)\, dv \qquad (2.34)$$

which is equivalent to the bandwidth of a flat, rectangular Fourier spectrum.

From the definition of the Fourier transform pair [Eqs. (2.16) and (2.17)], we see that Eqs. (2.33) and (2.34) can be written

$$\Delta t = \frac{F(0)}{f(0)} \qquad (2.35)$$

and

$$\Delta v = \frac{f(0)}{F(0)} \qquad (2.36)$$

Thus

$$\Delta t\, \Delta v = 1 \qquad (2.37)$$

which gives the lower bound condition of Eq. (2.28).

It should be noted that, if the symmetric pulse $f(t)$ contains *negative* values, as shown in Fig. 2.7, then the definitions of nominal duration and nominal bandwidth must be modified:

$$\Delta t |f(0)| \triangleq \int_{-\infty}^{\infty} |f(t)|\, dt \geq \left| \int_{-\infty}^{\infty} f(t)\, dt \right| = |F(0)| \qquad (2.38)$$

and

$$\Delta v |F(0)| \triangleq \int_{-\infty}^{\infty} |F(v)|\, dv \geq \left| \int_{-\infty}^{\infty} F(v)\, dv \right| = |f(0)| \qquad (2.39)$$

These definitions give rise to the uncertainty relation

$$\Delta t\, \Delta v \geq 1 \qquad (2.40)$$

which is essentially the condition of Eq. (2.28). From Fig. 2.7, we see that the nominal duration was determined by equating the area under the rectangular pulse function to the area under the curve of $|f(t)|$. It is evident that the nominal duration Δt is *wider* under this new definition of Eq. (2.38), provided $f(t)$ contains negative values. Similarly, the nominal bandwidth Δv of Eq. (2.39) can be interpreted in the same manner.

We now turn our attention to the inequality of Eq. (2.28) and show that it is essentially identical to *Heisenberg's uncertainty principle* in quantum mechanics [7]. In order to do so, let us denote by x a position variable for a particle, and by p the corresponding momentum variable. The uncertainty principle states that the position variable x and its momentum

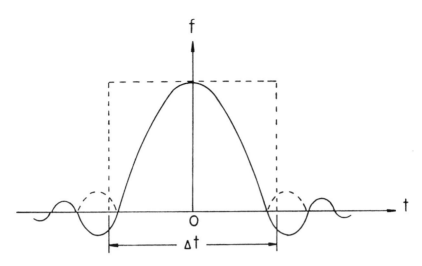

Fig. 2.7 Symmetric pulse with negative values.

variable p cannot be observed or measured simultaneously with arbitrary accuracy:

$$\Delta x \, \Delta p \geq h \tag{2.41}$$

where Δx and Δp are the position and momentum errors, respectively, and h is Planck's constant. The Heisenberg uncertainty relation of Eq. (2.41) can also be written in the form of energy and time variables:

$$\Delta E \, \Delta t \geq h \tag{2.42}$$

where ΔE and Δt are the corresponding energy and time errors. But since $E = h\nu$, hence $\Delta E = h \, \Delta \nu$, we see that

$$\Delta E \, \Delta t = h \, \Delta \nu \, \Delta t \geq h \tag{2.43}$$

Therefore one concludes that $\Delta \nu \, \Delta t \geq 1$ is in fact the *Heisenberg uncertainty relation*. The reader may realize how unrealistic it would be to select smaller values of $\Delta \nu$ and Δt, which would violate the uncertainty relation.

In concluding this section, we emphasize that time and frequency variables, as mentioned on several occasions, can be easily applied to *spatial coordinates*, namely, *spatial* and *spatial frequency* domains. For example, in most optical information processing [2], we use spatial domains, instead of the time variable.

Diffraction and Signal Analysis 49

2.5 DEGREES OF FREEDOM OF A SIGNAL

We now consider the problem of degrees of freedom of a band-limited signal. Let us denote by $f(t)$ a band-limited signal whose spectrum extends from zero frequency up to a definite maximum limit of v_m. We let $f(t)$ extend over a time interval of T, where $v_m T \gg 1$. Now the basic question is, How many sampling points or degrees of freedom are required in order to describe the function $f(t)$, over T, uniquely? To answer this fundamental question we present a simple example. However, before going into a detailed discussion, it is noted that $f(t)$ is not *completely* defined when only the value of $f(t)$, over T, is given. For simplicity, however, we let the value of $f(t)$ repeat for every interval of T:

$$f(t) = f(t+T) \tag{2.44}$$

which can be considered periodic for every T. Thus the function $f(t)$, over the period T, can be expanded in a Fourier series:

$$f(t) = \sum_{n=-M}^{M} C_n \exp(i2\pi n v_0 t) \tag{2.45}$$

where $v_0 = 1/T$, and $M = v_m T$.

From this Fourier expansion of Eq. (2.45), we see that $f(t)$ contains a finite number of terms:

$$N = 2M + 1 = 2v_m T + 1 \tag{2.46}$$

which includes the zero-frequency Fourier coefficient C_0. Thus, if the average value of $f(t)$, over T, is zero, then we see that $C_0 = 0$, and the total number of terms of the corresponding Fourier series reduces to

$$N = 2v_m T \quad \text{for } C_0 = 0 \tag{2.47}$$

If the duration T is sufficiently large, we see that Eq. (2.46) reduces to

$$N \simeq 2v_m T \tag{2.48}$$

which is the number of degrees of freedom required to specify $f(t)$, over T. In other words, it requires a total of N equidistant sampling points of $f(t)$, over T, to describe the function

$$t_s = \frac{T}{N} = \frac{1}{2v_m} \tag{2.49}$$

where t_s is the *Nyquist sampling interval*. Thus the corresponding *sampling frequency* is

$$f_s = \frac{1}{t_s} = 2v_m \tag{2.50}$$

which is called the *Nyquist sampling rate*. We see that

$$f_s \geq 2v_m \tag{2.51}$$

that is, the sampling frequency is at least equal to twice the highest frequency limit of $f(t)$. Now we consider reconstruction of the original function $f(t)$ with the corresponding N equidistant sample points, that is, the N degrees of freedom.

First, let us take the Fourier transform of $f(t)$:

$$F(v) = \int_{-\infty}^{\infty} f(t) \exp(-i2\pi v t)\, dt \tag{2.52}$$

By virtue of the band-limited nature of $f(t)$, we see that

$$F(v) = 0 \quad \text{for } v > v_m \tag{2.53}$$

Let $F(v)$ be made arbitrarily periodic in the frequency coordinates, as shown in Fig. 2.8, with a period of $2v_m$. Thus, in the frequency domain, $F(v)$ can be expanded in a Fourier series:

$$F(v) = \sum_{n=-\infty}^{\infty} K_n \exp\left(\frac{i\pi n v}{v_m}\right) \quad \text{for } |v| \leq v_m \tag{2.54}$$

and

$$F(v) = 0 \quad \text{for } |v| > v_m \tag{2.55}$$

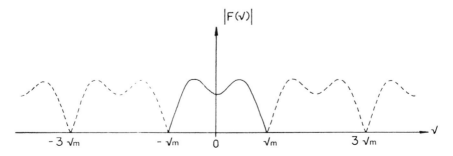

Fig. 2.8 Periodic representation of the Fourier spectrum.

with the corresponding Fourier coefficient K_n defined by

$$K_n = \int_{-v_m}^{v_m} F(v) \exp\left(-i\frac{\pi n v}{v_m}\right) dv \qquad (2.56)$$

But since $F(v)$ is the Fourier transform of $f(t)$, then $f(t)$ can be written

$$f(t) = \int_{-v_m}^{v_m} F(v) \exp(i 2\pi v t) \, dv \qquad (2.57)$$

In particular, at sampling points $t = -n/2v_m$, we have

$$f\left(-\frac{n}{2v_m}\right) = \int_{-v_m}^{v_m} F(v) \exp\left(-i\frac{\pi n v}{v_m}\right) dv = K_n \qquad (2.58)$$

which is equal to the Fourier coefficient K_n of Eq. (2.56). Thus we see that, $f(t)$ is given at various Nyquist intervals ($t = n/2v_m$), then the corresponding Fourier coefficient K_n can be obtained. But, from Eq. (2.54) we see that $F(v)$ can in turn be determined, and from Eq. (2.57) that knowledge of $F(v)$ implies a knowledge of $f(t)$. Therefore, if we substitute Eq. (2.54) into Eq. (2.57), we have

$$f(t) = \int_{-v_m}^{v_m} \sum_{n=-\infty}^{\infty} K_n \exp\left(\frac{i\pi n v}{v_m}\right) \exp(i 2\pi v t) \, dv \qquad (2.59)$$

By interchanging the integration and summation of Eq. (2.59), we obtain

$$f(t) = \sum_{n=-\infty}^{\infty} K_n \frac{\sin 2\pi v_m (t + n/2v_m)}{2\pi v_m (t + n/2v_m)} \qquad (2.60)$$

By substituting Eq. (2.58) into Eq. (2.60), we have

$$f(t) = \sum_{n=-\infty}^{\infty} f\left(\frac{n}{2v_m}\right) \frac{\sin 2\pi v_m (t - n/2v_m)}{2\pi v_m (t - n/2v_m)} \qquad (2.61)$$

where all the positive and negative signs of n have been adjusted.

From Eq. (2.61), we see that each sample of $f(t)$ is multiplied by a weighting factor $[(\sin x)/x]$, called a *sampling function*, that located every sample point of occurrence. This is in fact the output response of an ideal low-pass filter having a cutoff frequency at v_m (Fig. 2.6) when the samples $f(n/2v_m)$ are applied at the input end of the filter.

It is also noted that Shannon essentially used this concept, degrees of freedom, in the application of his sampling theory.

In concluding this section, we point out that the sampling theorem can be extended to higher-dimensional band-limited functions. For example, with a two-dimensional band-limited function $f(x, y)$, it can be shown that

$$f(x, y) = \sum_{n=-\infty}^{\infty} \sum_{m=-\infty}^{\infty} f(nx_0, my_0) \frac{\sin p_0(x - nx_0) \sin q_0(y - my_0)}{p_0(x - nx_0) q_0(y - my_0)}$$

(2.62)

where p_0 and q_0 are the highest angular frequencies with respect to the x and y coordinates of $f(x, y)$, and $x_0 = \pi/p_0$ and $y_0 = \pi/q_0$ are the respective Nyquist sampling intervals.

2.6 GABOR'S INFORMATION CELL

In 1946, Gabor [8] published a paper entitled "Theory of Communication" in the *Journal of the Institute of Electrical Engineers*. This was about 2 years before Shannon's [9] classical article, "A Mathematical Theory of Communication," appeared in the *Bell System Technical Journal*. Several of Gabor's concepts of information were quite consistent with Shannon's theory of information. In this section, we illustrate briefly one of his concepts, the information cell, which is related to the uncertainty principle in quantum mechanics. However, for a detailed discussion of Gabor [8] work on communication theory, we refer to his original papers.

Let us first take the frequency and the time variable to form a finite two-dimensional *Euclidean space*, as shown in Fig. 2.9, with v_m the maximum frequency limit and T the finite time sample of the signal function $f(t)$, This frequency-time space can be subdivided into elementary information elements or cells which Gabor called *logons*, such as

$$\Delta v \, \Delta t = 1 \tag{2.63}$$

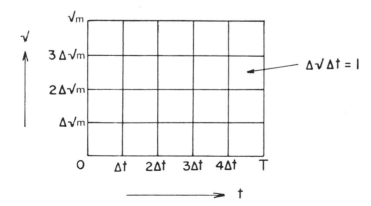

Fig. 2.9 Gabor's information cell.

Diffraction and Signal Analysis

We quickly recognize that Eq. (2.63) is essentially the lower bound of the uncertainty relation of Eq. (2.28). However, it is noted that the signal in each of the information cells has two possible *elementary signals*, a symmetric one and an *antisymmetric* one, with the same bandwidth Δv and the same duration Δt. Furthermore, the amplitudes of both these elementary signals should be given so that the signal function $f(t)$ can be uniquely described. From Fig. 2.9, we see that, for a given highest frequency content v_m and a finite time sample, the total number of information cells is

$$N_1 = v_m T \tag{2.64}$$

However, since each of the information cells contains two elementary functions, the total number of elementary functions is

$$N = 2N_1 = 2v_m T \tag{2.65}$$

which is the result of Eq. (2.48).

The *shapes* of the information cells are not particularly critical, but their unit area $\Delta v \Delta t = 1$, is. Moreover, from sampling technique, we see that the information cells are on the horizontal axis of the time coordinate and, as in Fourier analysis, the information cells are on the vertical axis of the frequency coordinate. For the elementary signals, Gabor [8, 10] suggested the use of *Gaussian cosine* and *Gaussian sine* signals, also called wavelets, as shown in Fig. 2.10.

It is seen from Eq. (2.63) that the elementary information cell suggested by Gabor is in fact the lower bound of the Heisenberg uncertainty principle in quantum mechanics:

$$\Delta E \Delta t = h \tag{2.66}$$

Since $E = hv$, we see that Eq. (2.66) is the same as Eq. (2.63).

It is emphasized that the band-limited signal must be a very special type of function; that is, the function must be well behaved. It contains no discontinuity or sharp angles and has only rounded-off features, as shown in Fig. 2.11. Thus this signal must be *analytic* over T.

In concluding this section, we note that the concept of Gabor's information cell can be applied to functions of a spatial domain. For example, given a spatial band-limited signal $f(x)$, the corresponding elementary information cell, in the angular spatial frequency p and spatial coordinate x, is

$$\frac{1}{2\pi} \Delta p \Delta x = 1 \tag{2.67}$$

The relation of Eq. (2.67) is used in calculating the spatial channel capacity in Chap. 3.

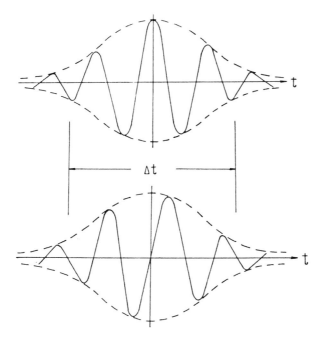

Fig. 2.10 The cosine and sine elementary signals, with Gaussian envelope.

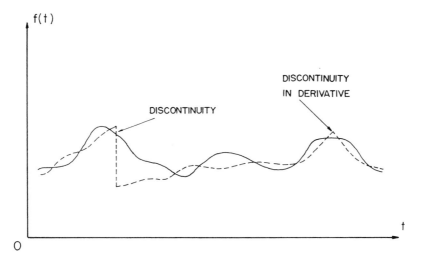

Fig. 2.11 Typical example of a band-limited signal.

2.7 SIGNAL DETECTION

A problem of considerable importance in optical communication and information processing is the detection of a signal embedded in random noise. Therefore, in this section, we discuss a special type of optimum linear filter known as a *matched filter*, which is remarkably useful in optical signal detection and in optical pattern or character recognition. We derive an expression for the filter transfer function on a somewhat general basis, namely, for a stationary additive noise.

It is well known in communication theory that the signal-to-noise ratio at the output end of a correlator can be improved to a certain extent. Let us consider the input excitation to a linear filtering system to be an additive mixture of a signal $s(t)$ and a stationary random noise $n(t)$:

$$f(t) = s(t) + n(t) \tag{2.68}$$

Let the output response of the linear filter due to the signal $s(t)$ alone be $s_0(t)$, and that due to the random noise $n(t)$ alone be $n_0(t)$. The figure of merit on which the filter design is based is the output signal-to-noise ratio at $t=0$:

$$\frac{S}{N} \triangleq \frac{|s_0(0)|^2}{\sigma^2} \tag{2.69}$$

where σ^2 is the mean-square value of the output noise.

In terms of the filter transfer function $H(v)$ and the Fourier transform $S(v)$ of the input signal $s(t)$, these quantities can be written

$$s_0(0) = \int_{-\infty}^{\infty} H(v) S(v)\, dv \tag{2.70}$$

and

$$\sigma^2 = \int_{-\infty}^{\infty} |H(v)|^2 N(v)\, dv \tag{2.71}$$

where $|H(v)|^2 N(v)$ is the power spectral density of the noise at the output end of the filter, and $N(v)$ is the power spectral density of the noise at the input end. Thus the output signal-to-noise ratio can be expressed explicitly in terms of the filter function $H(v)$:

$$\frac{S}{N} = \frac{\left| \int_{-\infty}^{\infty} H(v) S(v)\, dv \right|^2}{\int_{-\infty}^{\infty} |H(v)|^2 N(v)\, dv} \tag{2.72}$$

The objective of the filter design is to specify a filter function such that the output signal-to-noise ratio is maximum. To obtain such a filter transfer function, we can apply the *Schwarz inequality*, which states that

$$\frac{\left|\int_{-\infty}^{\infty} u(t)v^*(t)\,dt\right|^2}{\int_{-\infty}^{\infty} |u(t)|^2\,dt} \leq \int_{-\infty}^{\infty} |v^*(t)|^2\,dt \qquad (2.73)$$

where $u(t)$ and $v(t)$ are arbitrary functions, and the asterisk denotes the complex conjugate. The equality in Eq. (2.73) holds if and only if $u(t)$ is proportional to $v(t)$.

To make the Schwarz inequality applicable to Eq. (2.72), it may be expedient to express the output noise spectral density as the product of the two conjugate factors:

$$N(v) = N_1(v)N_1^*(v) \qquad (2.74)$$

Then Eq. (2.72) can be written

$$\frac{S}{N} = \frac{\left|\int_{-\infty}^{\infty} [H(v)N_1(v)][S^*(v)/N_1^*(v)]^*\,dv\right|^2}{\int_{-\infty}^{\infty} |H(v)N_1(v)|^2\,dv} \qquad (2.75)$$

If we identify the bracketed quantities of Eq. (2.75) as $u(t)$ and $v(t)$, in view of the Schwarz inequality, we then have

$$\frac{S}{N} \leq \int_{-\infty}^{\infty} \frac{|S(v)|^2}{N(v)}\,dv \qquad (2.76)$$

The equality in Eq. (2.76) holds if and only if the filter function is

$$H(v) = K\frac{S^*(v)}{N(v)} \qquad (2.77)$$

where K is a proportional constant. The corresponding value of the output signal-to-noise ratio is therefore

$$\frac{S}{N} = \int_{-\infty}^{\infty} \frac{|S(v)|^2}{N(v)}\,dv \qquad (2.78)$$

It is interesting to note that, if the stationary additive noise is white (i.e., the noise spectral density is uniform over all the frequency domain), then the optimum filter function is

$$H(v) = KS^*(v) \qquad (2.79)$$

Diffraction and Signal Analysis

which is proportional to the conjugate of the signal spectrum. This optimum filter is then said to be matched to the input signal $s(t)$. The output spectrum of the matched filter is therefore proportional to the power spectral density of the input signal:

$$G(v) = K|S(v)|^2 \tag{2.80}$$

Consequently, we see that the phase variation at the output end of the matched filter vanishes. In other words, the matched filter is capable of eliminating all the phase variations of $S(v)$ over the frequency domain.

2.8 STATISTICAL SIGNAL DETECTION

In Sec. 2.7 we illustrated that one of the objectives in signal filtering is to increase the signal-to-noise ratio. In fact, the increase in signal-to-noise ratio is purposely to minimize the probability of error in signal detection. It is noted, however, that in certain signal detections an increase in the signal-to-noise ratio does not necessarily guarantee minimizing the probability of error. But minimization of the probability of error can *always* be achieved by using the *decision process*. That is, the detection of signals can be achieved by making decisions about the presence or absence of the signals. This decision making gives rise to the basic concept of *threshold decision level*. Thus the decision must be made with respect to a threshold level chosen under the minimum probability of error criterion. We see that, if a decision is made that a signal was present but that it was caused by noise, then this event is called a *false alarm*. However, if a decision is made that no signal was present when a signal was actually present, then this event is called a *miss*. Therefore, for an optimum decision, the probability of error must be minimized; if the signal-to-noise ratio is high, *a priori*, then the threshold level can be raised higher, and the probability of error can be reduced. Thus the minimization of error probability can be achieved at a cost of high signal energy.

Let us consider the detection of *binary signals*. It is possible for us to establish a *Bayes' decision rule*. Through Bayes' *theorem*, the conditional probabilities have the relation

$$P(a/b) = \frac{P(a)P(b/a)}{P(b)} \tag{2.81}$$

One can write

$$\frac{P(a=0/b)}{P(a=1/b)} = \frac{P(a=0)P(b/a=0)}{P(a=1)P(b/a=1)} \tag{2.82}$$

where $P(a)$ is the *a priori* probability of a that is, $a=1$ corresponds to the signal presented, and $a=0$ corresponds to no signal.

We see that a logical decision rule is that, if $P(a=0/b)>P(a=1/b)$, then we decide that there is no signal ($a=0$) for a given b. However, if $P(a=0/b) < P(a=1/b)$, then we decide that there is a signal ($a=1$) for a given b. Thus, from Eq. (2.82), Bayes' decision rule can be written: Accept $a=0$ if

$$\frac{P(b/a=0)}{P(b/a=1)} > \frac{P(a=1)}{P(a=0)} \tag{2.83}$$

Accept $a=1$ if

$$\frac{P(b/a=0)}{P(b/a=1)} < \frac{P(a=1)}{P(a=0)} \tag{2.84}$$

It is clear that two possible errors can occur: if we accept that the received event b contains no signal ($a=0$), but the signal does in fact present ($a=1$) and, vice versa, if we accept that event b contains a signal ($a=1$), but the signal in fact is not present ($a=0$). In other words, the error of accepting $a=0$ when $a=1$ has actually occurred is a miss, and the error of accepting $a=1$ when $a=1$ has actually not occurred is a false alarm.

Let us assign the cost values C_{00}, C_{01}, C_{10}, and C_{11}, respectively to the following cases: (1) $a=0$ is actually true, and the decision is to accept it; (2) $a=0$ is actually true, and the decision is to reject it; (3) $a=1$ is actually true, and the decision is to reject it; and (4) $a=1$ is actually true, and the decision is to accept it. Thus the overall *average cost* is

$$\bar{C} = \sum_{i=0}^{1} \sum_{j=0}^{1} C_{ij} P(a_i) P(B_j/a_i) \tag{2.85}$$

where $a_0 = 0$, $a_1 = 1$, $P(a_i)$ is the *a priori* probability of a_i, and $P(B_j/a_i)$ is the conditional probability that b falls in B_j if a_i is actually true, as shown in Fig. 2.12.

To minimize the average cost, it is desirable to select a region B_0, where $B_1 = B - B_0$, in such a manner that the average cost is minimum, in other words, to place certain restrictions on the cost values C_{ij} so that $\bar{C}$ will be minimum for a desirable B_0. For example, a miss or a false alarm may be costlier than correct decisions:

$$C_{10} > C_{11} \tag{2.86}$$

and

$$C_{01} > C_{00} \tag{2.87}$$

Diffraction and Signal Analysis 59

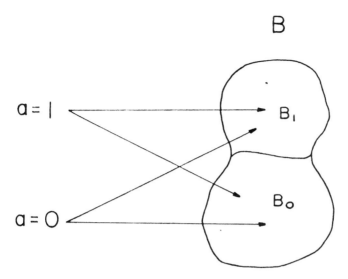

Fig. 2.12 Hypothesis of the received events for two possible transmitted events.

But since

$$P(B_j/a_i) = \int_{B_j} p(b/a_i)\, db \qquad i,j = 0, 1 \tag{2.88}$$

and

$$P(B_1/a_i) = 1 - P(B_0/a_i) \qquad i = 0, 1 \tag{2.89}$$

by substituting Eqs. (2.86) and (2.87) into Eq. (2.85), we have

$$\bar{C} = C_{01}P(a=0) + C_{11}P(a=1) + \int_{B_0} P(a=1)(C_{10} - C_{11})p(b/a=1)\, db$$
$$- \int_{B_0} P(a=0)(C_{01} - C_{00})p(b/a=0)\, db \tag{2.90}$$

where the first and second terms are positive constants.

In the minimization of $\bar{C}$, we select an optimum region of B_0. But because of Eqs. (2.86) and (2.87) we see that it is sufficient to select region B_0 such that the second integral of Eq. (2.90) is larger than the first integral. Thus equivalently,

$$\frac{P(b/a=0)}{P(b/a=1)} > \frac{P(a=1)(C_{10} - C_{11})}{P(a=0)(C_{01} - C_{00})} \tag{2.91}$$

Now let us write

$$\alpha \triangleq \frac{P(b/a=0)}{P(b/a=1)} \tag{2.92}$$

which is the *likelihood ratio*, and

$$\beta \triangleq \frac{P(a=1)(C_{10}-C_{11})}{P(a=0)(C_{01}-C_{00})} \tag{2.93}$$

which is simply a constant incorporated with the *a priori* probabilities and the error costs. The decision rule is to select the hypothesis for which the signal is actually absent, if $\alpha > \beta$.

If the inequality of Eq. (2.91) is reversed ($\alpha < \beta$), then one chooses B_1 instead. In other words, Bayes' decision rule [Eq. (2.91)] ensures a minimum average cost for the decision making.

Furthermore, if the costs of the errors are equal, $C_{10} = C_{01}$, then the decision rule reduces to Eqs. (2.83) and (2.84). If the decision making has sufficient information on the error costs, then one uses Bayes' decision rule of Eq. (2.91) to begin with. However, if information on the error costs is not provided, then one uses the decision rule as given by Eqs. (2.83) and (2.84).

It is also noted that the Bayesian decision process depends on the *a priori* probabilities $P(a=0)$ and $P(a=1)$. However, if the *a priori* probabilities are not provided but one wishes to proceed with decision making alone, then the likelihood ratio test can be applied. That is, if

$$\frac{P(b/a=0)}{P(b/a=1)} > 1 \tag{2.94}$$

then one accepts $a=0$ for the received event of b. But, if

$$\frac{P(b/a=0)}{P(b/a=1)} < 1 \tag{2.95}$$

then one accepts $a=1$.

From Eq. (2.91), we see that Eq. (2.94) implies that

$$P(a=1)(C_{10}-C_{11}) = P(a=0)(C_{01}-C_{00})$$

Thus, if $C_{10} - C_{11} = C_{01} - C_{00}$, then the *a priori* probabilities of $P(a)$ are equal. It is noted that applications of the likelihood ratio test are limited, since the decision making takes places without knowledge of the *a priori* probabilities. Thus the results frequently are quite different from those from the minimum-error criterion.

Although there are times when there is no satisfactory way to assign appropriate *a priori* probabilities and error costs, there is a way of

Diffraction and Signal Analysis 61

establishing an optimum decision criterion, namely, the *Neyman-Pearson criterion* [11]. One allows a fixed false-alarm probability and then makes the decision in such a way as to minimize the miss probability.

Since it is not our present objective to discuss this subject in detail, we refer interested readers to the texts by Middleton [12], Selin [13], and Van Trees [14].

2.9 SIGNAL RECOVERING

An interesting problem in optical processing must be signal recovering (e.g., image restoration). The type of optimum filter is the solution based on minimization of the mean-square error [15], as given by

$$\min \langle \overline{\varepsilon^2(t)} \rangle = \min \langle \lim_{T \to \infty} \frac{1}{2T} \int_{-\infty}^{\infty} [f_o(t) - f_d(t)]^2 \, dt \rangle \quad (2.96)$$

where $f_o(t)$ and $f_d(t)$ are the actual output and the desired output response of the filter. A linear filter that is synthesized by using minimum-mean-square error criterion is known as the *optimum linear filter*.

Let us look at the mean-square-error performance of a linear filter function as given by

$$\overline{\varepsilon^2(t)} = \lim_{T \to \infty} \frac{1}{2T} \int_{-T}^{T} \left[\int_{-\infty}^{\infty} h(\tau) f_i(t - \tau) \, d\tau - f_d(t) \right]^2 dt \quad (2.97)$$

where $h(t)$ is impulse response of the filter and $f_i(t)$ is the input signal function. By expanding the preceding equation, it can be shown that the mean-square error can be written as

$$\overline{\varepsilon^2(t)} = \int_{-\infty}^{\infty} h(\tau) \, d\tau \int_{-\infty}^{\infty} h(\sigma) \, d\sigma R_{ii}(\tau - \sigma) - 2 \int_{-\infty}^{\infty} h(\tau) \, d\tau R_{id}(\tau) + R_{dd}(0)$$

$$(2.98)$$

where $h(t)$ is the impulse response of the filter, $R_{ii}(\tau)$ is the autocorrelation function of the input signal, $R_{id}(\tau)$ is the crosscorrelation of the input with the desired-output function, and $R_{dd}(0)$ is the mean square value of the desired-output function, as given by

$$R_{ii}(\tau) = \int_{-\infty}^{\infty} f_i(t) f_i(t + \tau) \, dt \quad (2.99)$$

$$R_{\text{id}}(\tau) = \int_{-\infty}^{\infty} f_i(t) f_d(t+\tau)\, dt \tag{2.100}$$

$$R_{\text{dd}}(\tau) = \int_{-\infty}^{\infty} f_d(t) f_d(t+\tau)\, dt \tag{2.101}$$

$$R_{\text{dd}}(0) = \int_{-\infty}^{\infty} [f_d^2(t)]\, dt \tag{2.102}$$

It is trivial to see that the mean-square error is a function of $h(t)$, which is the center problem to determine $h(t')$ by minimizing the error. Since the minimization of functionals is a subject to *calculus of variation*, which leads to the *Wiener-Hopf* integral equation as given by

$$\int_{-\infty}^{\infty} h(\sigma) R_{\text{ii}}(\tau - \sigma)\, d\sigma - R_{\text{id}}(\tau) = 0 \tag{2.103}$$

We note that by following the realizable constrain of a physical filter, strictly speaking the preceding result should be written as

$$\int_{-\infty}^{\infty} h(\sigma) R_{\text{ii}}(\tau - \sigma) d\sigma - R_{\text{id}}(\tau) \neq 0 \qquad \text{for } \tau < 0 \tag{2.104}$$

$$\int_{-\infty}^{\infty} h(\sigma) R_{\text{ii}}(\tau - \sigma) d\sigma - R_{\text{id}}(\tau) = 0 \qquad \text{for } \tau \geq 0 \tag{2.105}$$

An interest point of the Wiener-Hopf intergral equation is in fact corresponds to *Euler-Lagrange* differential equation in the calculus of variation. To further emphasize the Wiener-Hopf equation such as

$$R_{\text{id}}(\tau) = \int_{-\infty}^{\infty} h_{\text{opt}}(\tau) R_{\text{ii}}(\tau - \sigma)\, d\sigma \tag{2.106}$$

for which the optimum filter $h_{\text{opt}}(\tau)$ is needed to solve.

If the desired output could be obtained by assuming without the physical realizable constrain, the input-output crosscorrelation of the optimum filter is, in fact, the input and the desired-output cross-correlation itself, which holds not only for $\tau \geq 0$ but also for $\tau < 0$. Now the problem becomes a trivial one; by taking the Fourier transform of Eq. (2.106), we have

$$R_{\text{id}}(v) = H_{\text{opt}}(v) R_{\text{ii}}(v) \tag{2.107}$$

Diffraction and Signal Analysis 63

where $H_{\text{opt}}(v)$ is the transform function of the optimum filter, which can be written as

$$H_{\text{opt}}(v) = \frac{R_{\text{id}}(\omega)}{R_{\text{ii}}(\omega)} \tag{2.108}$$

That is in fact the solution of our problem.

Mention must be made that we do not exactly know the desired output response, but we have been demanding the best filter by minimizing the mean-square error, by which gives us the *best* approximation. In other words, this method is trying to get the best result we can, similar to bargaining with a used-car salesman. As we demanded more than a filter can be synthesized, the input-output cross-correlation *cannot* be equal to the crosscorrelation of the input and the desired output. According to the Wiener-Hopf equation, the filter design should be based on the minimum error criterion and the impulse response should be compelled to the inequality of $\tau \geq 0$, but allow it to vary somewhat for $\tau < 0$, such as

$$q(\tau) = R_{\text{id}}(\tau) - \int_{-\infty}^{\infty} h_{\text{opt}}(\sigma) R_{\text{ii}}(\tau - \sigma) \, d\sigma \tag{2.109}$$

where $q(\tau) \neq 0$ for $\tau < 0$ but vanishes for $\tau \geq 0$.

For the readers who are interested in the Wiener-Hopf solution, see the text by Lee [15].

2.10 SIGNAL AMBIGUITY

The choice of a transmitted waveform for a radar seems more trivial than it is for communication. However, it is actually not so fundamental, as it does not contain any information. Nevertheless, it is the reflected (or echo) wave from the target that the temporal and the doppler-shift (i.e., frequency) provide the range and radial velocity information of the target. Whenever time and frequency are mentioned, we would anticipate to relate the quantities to the uncertainty relationship. In other words, no wave form can occupy a very short time and having a very narrow bandwidth. Thus, there is a limitation imposed on the combined resolution in time and frequency [16].

To simplify the analysis, we set scaler wave field $u(t)$ be constrained by the following equation:

$$\int |u(t)|^2 \, dt = \int |U(v)|^2 \, dv = 1 \tag{2.110}$$

In view of the Wiener-Khinchin and Parseval's theorems, we see that in the time domain we have

$$R(\tau) = \int u(t)u^*(t+\tau)\,dt = \int |U(v)|^2 \, e^{i2\pi v\tau}\,dv \tag{2.111}$$

and

$$T = \int |R(\tau)|^2\,d\tau = \int |u(v)|^4\,dv$$

where the super asterisk denotes the complex conjugate. Similarly in frequency domain, we have

$$K(v) = \int U^*(v')U(v'+v)\,dv' = \int |u(t)|^2 \, e^{-i2\pi vt}\,dt \tag{2.112}$$

and

$$F = \int |K(v)|^2\,dv = \int |u(t)|^4\,dt \tag{2.113}$$

where F is defined as the frequency resolution cell, which provides the well-known uncertainty relationship as given by [see Eg. (2.63)]

$$(\text{Frequency resolution cell}) \times (\text{time span}) = 1 \tag{2.114}$$

Notice that Eqs. (2.111) applies frequently to the resolution of detecting stationary targets at different ranges, while Eqs. (2.112) applies to the resolution of detecting moving targets at different radial velocities. However, when the targets are both at different range, and moving with different radial velocities, neither quantity being known in advance, the separation of time (e.g., range) and frequency (i.e., radial velocity) resolution do not always provide the actual resolving power of the target. Thus, we need a more general description of time and frequency shift to interpret the resolving power of the signal, which is known as the *ambiguity function*, as given by

$$\chi(\tau, v) = \int u(t)u^*(t+\tau)\,e^{-i2\pi vt}\,dt = \int U^*(v')U(v'+v)\,e^{i2\pi v'\tau}\,dv' \tag{2.115}$$

The significance of $\chi(\tau, v)$ is that the range and (radial) velocity of a target cannot be resolved at $(\tau_0 + \tau, v_0 + v)$, where τ_0 and v_0 are the mean range and velocity of the target. We note that $\chi(\tau, v)$ is equal to unity, i.e.,

$$\iint |\chi(\tau, v)|^2 \, d\tau\,dv = 1 \tag{2.116}$$

which is called the *area of ambiguity* in the time-frequency domain.

Diffraction and Signal Analysis

Example

One of the simplest example of ambiguity function is obtained from a single Gaussian pulse as given by

$$\mu(t) = \sqrt[4]{2}\, e^{-\pi t^2}$$

By substituting into Eq. (2.115), we have

$$\chi(\tau, v) = \exp\left(-\frac{1}{2}\pi\tau^2\right) \exp\left(-\frac{1}{2}\pi v^2\right) \exp(i\pi v\tau)$$

From the preceding equation we see that the magnitude of ambiguity distribution, as given by

$$|\chi(\tau, v)|^2 = e^{-\pi(\tau^2 + v^2)}$$

describes a circular pattern in the (τ, v) domain, as can be seen in Fig. 2.13. If the pulse is shortened, it can be seen that the circular ambiguity pattern would become elliptical, as illustrated in Fig. 2.14.

A less trivial example is that $u(t)$ represents a train of Gaussian pulses, modulated by a broad Gaussian envelope as shown in Fig. 2.15a. The corresponding ambiguity $|\chi(\tau, v)|^2$ is sketched in Fig. 2.15b, in which the ambiguity in frequency resolution is split up into narrow bands, whereas in time it is repeated itself. Notice that the shaded areas of the redistributed ambiguity are in fact equal to the area of the single pulse ambiguity, i.e.,

$$\iint |\chi(\tau, v)|^2 \, d\tau \, dv = 1 \qquad (2.117)$$

Mention must be made that in actual application, it is not our interest in resolving all the ranges of the target for which the redundant parts of the ambiguity diagram remote from the origin have no practical usage.

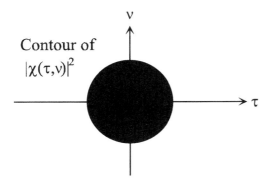

Fig. 2.13 Ambiguity of a Gaussian pulse.

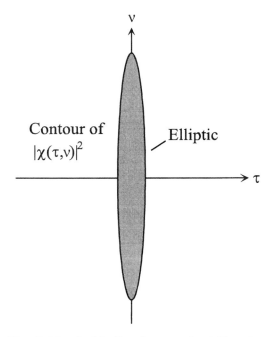

Fig. 2.14 Ambiguity of a very short Gaussian pulse.

For example, a target at 10 mi away should not be confused at 1010 mi. Therefore, in practice, we may draw a rectangular region around the origin of the (τ, v) coordinate plane to discriminate the ambiguity outside the region. Thus, we see that the pulse train example will be a more acceptable waveform for range and (radial) velocity discrimination, as compared with the ambiguity distribution obtained from single pulse. However, when interpreting the ambiguity diagram, one should remember that τ and v do not represent the actual range and (radial) velocity, but rather the difference between the ranges and velocities of any two targets that need to be resolved.

2.11 WIGNER SIGNAL REPRESENTATION

There is another form of time-frequency signal representation, as similar to ambiguity function, is defined by [17,18]

$$W(\tau, v) = \int u(t) u^*(\tau - t) e^{-i2\pi vt} \, dt = \int U^*(v') U(v - v') e^{i2\pi v' \tau} \, dv'$$

(2.118)

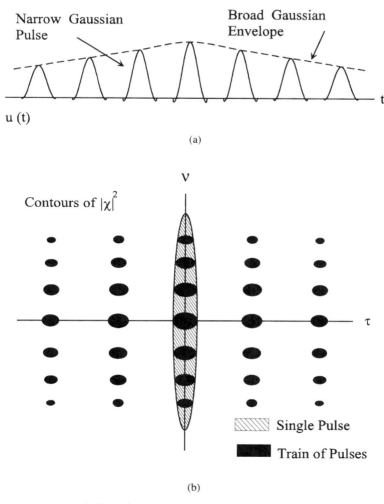

Fig. 2.15 (a) A Gaussian pulse-train. (b) Ambiguity diagram of (a).

which is known as the *Wigner distribution function*. Instead of using correlation operator [i.e., $u(t)u^*(t+\tau)$], Wigner used the convolution operator [i.e., $u(t)u^*(\tau - t)$] for his transformation. The physical properties of the Wigner distribution function (WDF) can be shown as

$$\int W(\tau, v)\,dv = |u(v)|^2 \qquad (2.119)$$

$$\int W(\tau, v) \, d\tau = |U(v)|^2 \tag{2.120}$$

$$\int W(\tau, v) \, d\tau \, dv = 1 \tag{2.121}$$

in which the WDF has been normalized to unity for simplicity.

One of the interesting features of WDF is the inversion. Apart the association with a constant phase factor, the transformation is unique, as can be shown:

$$\int W(\tau, v) e^{-i4\pi v\tau} \, d\tau = U(2v) U^*(0) \tag{2.122}$$

$$\int W(\tau, v) e^{i4\pi v\tau} \, dv = u(2\tau) u^*(0) \tag{2.123}$$

If the constant phase factors $u^*(0)$ or $U^*(0)$ happen to be very small or even zero, then we can reformulate the transformations by using the maximum value of $|u(\tau)|^2$ or $|U(\tau)|^2$, which occurs at $\tau_{\max}$, or $f_{\max}$, respectively, as given by

$$\int W(\tau, v) \exp\left[-4\pi(v - v_{\max})\tau\right] d\tau = U(2f - v_{\max}) U^*(v_{\max}) \tag{2.124}$$

$$\int W(\tau, v) \exp\left[i4\pi v(\tau - \tau_{\max})\right] dv = u(2\tau - \tau_{\max}) u^*(\tau_{\max}) \tag{2.125}$$

In other words, WDF does provide the frequency content of a time varying signal, instead of providing the time-frequency resolution as the ambiguity function does.

Examples

In the following, we provide a few simple examples of the WDF. First let us consider a single-tone signal as given by

$$u(t) = e^{i2\pi v_0 t}$$

By substition into Eq. (2.118), we have

$$W(\tau, v) = \delta(v - v_0)$$

which is shown in Fig. 2.16a.

If the signal is a single-pulse function, such as

$$u(t) = \delta(t - t_0)$$

Diffraction and Signal Analysis 69

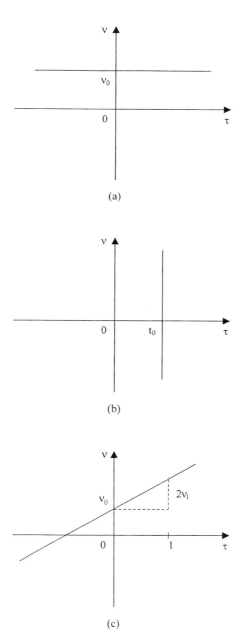

Fig. 2.16 Samples of Wigner distributions (a) for a single tone signal, (b) for a pulse signal, (c) for a chirp signal.

then the corresponding WDF would be

$$W(\tau, v) = \delta(\tau - t_0)$$

which is plotted in Fig. 2.16b.

On the other hand, if it is a chirp signal as given by

$$u(t) = \exp(i2\pi v_1 t^2 + v_0 t)$$

then the corresponding WDF can be shown as

$$W(\tau, v) = \delta(v - 2v_1\tau - v_0)$$

as illustrated in Fig. 2.16c.

2.12 FOURIER TRANSFORM PROPERTIES OF LENSES

It is extremely useful to obtain a two-dimensional Fourier transformation from a positive lens [19, 20]. Fourier transform operations usually bring to mind complicated electronic spectrum analyzers or digital computers. However, this complicated transformation can be performed extremely simply in a coherent optical system; and because the optical transform is two-dimensional, it has a greater information capacity than transforms carried out by means of electronic circuitry.

In the derivation of the Fourier transform properties of a lens, we call attention to the Fresnel-Kirchhoff theory discussed in Sec. 2.2. The complex light field on a planar surface (α, β) due to a point source at (x, y) plane, as shown in Fig. 2.4, can be described by

$$h_l(\boldsymbol{\sigma} - \boldsymbol{\rho}) = C \exp\left\{i\frac{\pi}{\lambda l}[(\alpha - x)^2 + (\beta - y)^2]\right\} \tag{2.126}$$

where $C = \exp(ikl)/l$ is a complex constant. Equation (2.126) is known as a *spatial impulse response* (Sec. 2.2).

Thus the complex light field due to a coherent source $f(x, y)$ can be written as

$$g(\alpha, \beta) = \int\int_{-\infty}^{\infty} f(x, y) h_l(\boldsymbol{\sigma} - \boldsymbol{\rho}) \, dx \, dy \tag{2.127}$$

which is, equivalently,

$$g(\alpha, \beta) = f(x, y) * h_l(\boldsymbol{\sigma} - \boldsymbol{\rho}) \, dx \, dy \tag{2.128}$$

where the asterisk denotes the convolution operation.

However, for Fourier transformation, a lens is needed for the operation. Now let us consider Fig. 2.17, a simple optical system. If the

Diffraction and Signal Analysis 71

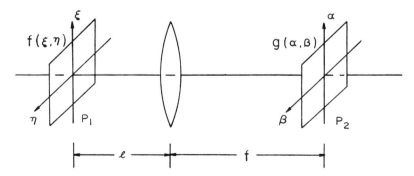

Fig. 2.17 Geometry for determination of the optical Fourier transformation.

complex light field at P_1 is $f(\xi, \eta)$, then the complex light distribution at P_2 can be written

$$g(\alpha, \beta) = C\{[f(\xi, \eta) * h_l(\xi, \eta)]T(x, y)\} * h_f(x, y) \qquad (2.129)$$

where C is an arbitrary complex constant, $h_l(\xi, \eta)$ and $h_f(x, y)$ are the corresponding spatial impulse responses, and $T(x, y)$ is the *phase transformation* of the lens. Equation (2.129) can be written in the integral form:

$$g(\alpha, \beta) = C \iint_{S_1} \left[\iint_{S_2} \exp\left(i\frac{k}{2}\Delta\right) dx\, dy \right] f(\xi, \eta)\, d\xi\, d\eta \qquad (2.130)$$

where S_1 and S_2 denote the surface integrals of the light field P_1 and the lens T, respectively, and

$$\Delta = \left\{\frac{1}{l}[(x-\xi)^2 + (y-\eta)^2] + \frac{1}{f}[(\alpha-x)^2 + (\beta-y)^2 - (x^2-y^2)]\right\}$$
$$(2.131)$$

It is noted that the surface integral of the lens may be assumed to be of infinite extent, since the lens is very large compared to the spatial apertures at P_1 and P_2 paraxiality.

Equation (2.131) can be written

$$\Delta = \frac{1}{f}(\mu\xi^2 + \mu x^2 + \alpha^2 - 2\mu\xi x - 2x\alpha - \mu\eta^2 + \mu y^2 + \beta^2 - 2\mu\eta y - 2y\beta)$$
(2.132)

where $\mu = f/l$.

By completing the square, Eq. (2.132) can be written

$$\Delta = \frac{1}{f}[\mu^{1/2}x - \mu^{1/2}\xi - \mu^{-1/2}\alpha)^2 - \alpha^2 \frac{1-\mu}{\mu} 2\xi\alpha$$
$$+ (\mu^{1/2}y - \mu^{1/2}\eta - \mu^{-1/2}\beta)^2 - \beta^2 \frac{1-\mu}{\mu} - 2\eta\beta]$$
(2.133)

By substituting in Eq. (2.130), we have

$$g(\alpha, \beta) = C \exp\left[-i\frac{k}{2f}\left(\frac{1-\mu}{\mu}\right)(\alpha^2 + \beta^2)\right]$$

$$\cdot \iint_{S_1} f(\xi, \eta) \exp\left[-i\frac{k}{f}(\alpha\xi + \beta\eta)\right] d\xi\, d\eta$$
(2.134)

$$\cdot \iint_{S_2} \exp\left\{i\frac{k}{2f}[(\mu^{1/2}x - \mu^{1/2}\xi - \mu^{-1/2}\alpha)^2\right.$$
$$\left. + (\mu^{1/2}y - \mu^{1/2}\eta - \mu^{-1/2}\beta)^2]\right\} dx\, dy$$

Since the integrations over S_2 are assumed to be taken from $-\infty$ to ∞, we obtain a complex constant which can be incorporated with C. Thus we have

$$g(\alpha, \beta) = C \exp\left[-i\frac{k}{2f}\left(\frac{1-\mu}{\mu}\right)(\alpha^2 + \beta^2)\right]$$

$$\cdot \iint_{S_1} f(\xi, \eta) \exp\left[-i\frac{k}{f}(\alpha\xi + \beta\eta)\right] d\xi\, d\eta$$
(2.135)

From this result we see that, except for a spatial quadratic phase variation, $g(\alpha, \beta)$ is the Fourier transform of $f(\xi, \eta)$. In fact, the quadratic phase factor vanishes for $l = f$. Evidently, if the signal plane P_1 is placed at the front focal

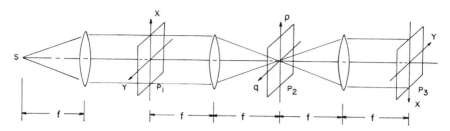

Fig. 2.18 Successive Fourier transformations of lenses. A monochromatic point source is located at S.

plane of the lens, the quadratic phase factor disappears, which leaves an exact Fourier transform relation. Thus Eq. (2.135) can be written

$$G(p, q) = C_1 \iint_{S_1} f(\xi, \eta) \exp[-i(p\xi + q\eta)] d\xi, d\eta \quad \text{for } \mu = 1$$

(2.136)

where $p = k\alpha/f$ and $q = k\beta/f$ are the spatial frequency coordinates.

It is emphasized that the exact Fourier transform relation takes place under the conditions $l = f$. Under the conditions $l \neq f$, a quadratic phase factor is included. Furthermore, it can easily be shown that a quadratic phase factor also results if the signal plane P_1 is placed behind the lens.

In conventional Fourier transform theory, transformation from the spatial domain to the spatial frequency domain requires the kernel exp $[-i(px + pq)]$, and transformation from the spatial frequency domain to the spatial domain requires the conjugate kernel $\exp[i(px + qy)]$. Obviously, a positive lens always introduces the kernel $\exp[-i(px + qy)]$. Since an optical system takes only successive transforms, to perform the inverse transform one can simply reverse the coordinates axes at the output plane, as shown in Fig. 2.18.

REFERENCES

1. M. Born and E. Wolf, *Principle of Optics*, 2nd rev. ed., Pergamon, New York, 1964.
2. F. T. S. Yu, *Introduction to Diffraction, Information Processing, and Holography*, MIT Press, Cambridge, Mass., 1973.
3. W. B. Davenport, Jr., and W. L. Root, *Random Signals and Noise*, McGraw-Hill, New York, 1958.

4. A. Papoulis, *The Fourier Integral and Its Applications*, McGraw-Hill, New York, 1962.
5. R. N. Bracewell, *The Fourier Transform and Its Applications*, McGraw-Hill, New York, 1965.
6. E. A. Guillemin, *Synthesis of Passive Networks*, John Wiley, New York, 1957.
7. J. I. Powell and B. Crasemann, *Quantum Mechanics*, Addison-Wesley, Reading, Mass., 1961.
8. D. Gabor, "Theory of Communication," *J. Inst. Electr. Eng.*, vol. 93, 429 (1946).
9. C. E. Shannon, "A Mathematical Theory of Communication," *Bell Syst. Tech. J.*, vol. 27, 379, 623 (1948).
10. D. Gabor, "Communication Theory and Physics," *Phil. Mag.*, vol. 41, no. 7, 1161 (1950).
11. J. Neyman and E. S. Pearson, "On the Use and Interpretation of Certain Test Criteria for Purposes of Statistical Inference," *Biometrika*, vol. 20A, 175, 263 (1928).
12. D. Middleton, *Statistical Communication Theory*, McGraw-Hill, New York, 1960.
13. I. Selin, *Detection Theory*, Princeton University Press, Princeton, N.J., 1965.
14. H. L. Van Trees, *Detection, Estimation, and Modulation Theory*, John Wiley, New York, 1968.
15. Y. W. Lee, "Statistical Theory of Communication", Wiley, New York, 1961.
16. P. M. Woodward, *Probability and Information Theory, with Application to Radar*, Pergamon Press, London, 1953.
17. E. Wigner, "On the quantum correction for thermodynamic equilibrium," *Phys Rev.* vol. 40, 49–759 (1932).
18. H. O. Bartelt, K. H. Brenner, and A. W. Lohmann, "The Wigner Distribution Function and Its Optical Production," *Opt. Commun.*, vol. 32, 32 (1980).
19. L. J. Cutrono et al., "Optical Data Processing and Filtering Systems," *IRE Trans. Inform. Theory*, IT-6, 386 (1960).
20. F. T. S. Yu and S. Jutamulia, *Optical Signal Processing, Computing and Neural Networks*, Wieley-Interscience, New York, 1992.

3
Optical Spatial Channel and Encoding Principles

In Chap. 1 we defined the basic concept of information theory: If H is the average information per unit resolution cell provided by an optical source, and if the source is emitting at a rate of m resolution cells per second, then the rate of information emitted by the optical source is

$$I = mH \quad \text{bits/sec} \tag{3.1}$$

It is noted that in most of the optical information source there is some degree of *temporal* and *spatial* redundancy. To reduce this redundancy it is convenient to replace the original signal (or image) by a different set of new signal elements, that is, optical or image codes. The general procedure for replacing original signals is to perform a certain encoding operation. If the decoding procedure at the receiving end is assumed to be unique, then the encoding and decoding processes will result in no loss of information. However, if the decoding is not a unique process, then some information will be lost. Thus one of the required restrictions on the basic coding theory is a uniquely decodable code.

It can be seen that numerous uniquely decodable codes exist. However, we consider the set of uniquely decodable codes with which the highest efficiency in information transmission can be achieved. There is no general method of obtaining the most efficient code, since coding depends on the intention to transfer the original signals. In this connection, we center our attention on the optical channel in which the encoded signals (or images) are transmitted or stored.

3.1 OPTICAL SPATIAL COMMUNICATION CHANNEL

In general, an optical channel may be considered a space- and time-variable channel. The channel may be described as a function of spatial coordinates which vary with respect to time, that is, $f(x, y; t)$, where x and y are the orthogonal spatial coordinates. In the following we discuss the optical channel separately with respect to spatial and temporal domains. General optical channel capacity with respect to this concept is also treated.

If it is assumed that we want the encoded spatial signal (the image) to be transmitted through an optical spatial channel or to be recorded (stored) on a photosensitive plate (photographic plate), then we will search for the smallest spatial bandwidth Δp of the optical spatial channel. The spatial signal transmitted through the channel with a resolution element Δx is

$$\frac{1}{2\pi}\Delta p \, \Delta x = 1 \qquad (3.2)$$

where Δp is the spatial bandwidth in radians per unit distance, and Δx is the resolution element in distance.

It is noted that the spatial constraints of Eq. (3.2) may not be sufficient to specify a general optical channel, since a physical channel is usually restricted by a certain time imposition. It is the time constraint that must also be considered. If it is assumed that the optical spatial channel possesses a system bandwidth of $\Delta \omega$ in radians per second, then the shortest time interval Δt required for the channel to respond efficiently will be

$$\Delta t \geq \frac{2\pi}{\Delta \omega} \qquad (3.3)$$

Equation (3.3) implies that the smallest bandwidth required for the optical channel for a resolution time interval Δt to be effectively transmitted is

$$\frac{1}{2\pi}\Delta \omega \, \Delta t = 1 \qquad (3.4)$$

Now let us consider a set of resolution cells in the spatial domain of the channel. For simplicity, we assume there is an $m \times n$ array of resolution cells, as shown in Fig. 3.1. And we let each cell be centered at each (x_i, y_j), where $i = 1, 2, \ldots, m$, and $j = 1, 2, \ldots, n$. We also select a set of signal levels (i.e., different levels of irradiance) for the optical spatial communication. Once we have adopted these sets of cells and signal levels, an optical spatial communication channel can be defined.

Based on restrictions of Eq. (3.2), we can select the set of resolution cells and, with the constraints of Eq. (3.4), we can consider statistically a set of signal levels quantized into k discrete irradiances, $E_1, E_2, \ldots, E_k$. If we assume that the incremental time duration of the spatial signal is Δt, then we can use it to make the random selection of each of the signal levels within a cell, Thus the optical spatial channel can be defined, within

Optical Spatial Channel and Encoding Principles 77

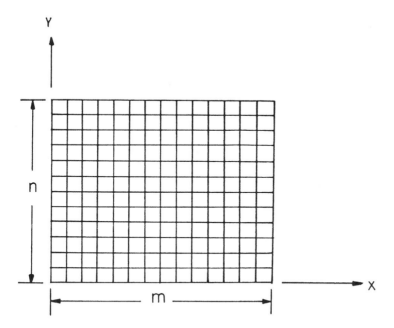

Fig. 3.1 An $m \times n$ array of resolution cells.

the time interval Δt, by the $m \times n$ array of resolution cells and the different signal levels E_i.

Now if we assume that the E_i's, within the resolution cells, distributed in the spatial domain are statistically independent, then an arbitrary distribution of the signal levels across the $m \times n$ array of resolution cells can represent a spatially coded message. Thus, for a number of $m \times n$ arrays, a spatially coded message is simply the combination of the individual signal levels E_i over the $m \times n$ array of cells.

Now let N be the total number of these combinations. If these N combinations are assumed, *a priori*, to be equiprobable, then the information provided by each of the spatially coded messages (i.e., the combination of E_i over $m \times n$ arrays) is

$$I = \log_2 N \quad \text{bits/message} \tag{3.5}$$

But since the total number of combinations can be written

$$N = k^{mn} \tag{3.6}$$

the information provided by each of the spatially coded messages can be written

$$I = mn \log_2 k \qquad \text{bits/message} \tag{3.7}$$

The corresponding rate of information transmission through the optical spatial channel is therefore

$$R = \frac{I}{\Delta t} = \frac{mn}{\Delta t} \log_2 k \qquad \text{bits/sec} \tag{3.8}$$

It is noted that, if the arrays of $m \times n$ resolution cells are not the same size, then the total combinations of the cells N, strictly speaking, is a function over all the spatial plane S, that is, $N(S)$. If we let $\sigma_1, \sigma_2, \ldots, \sigma_q$ be the spatial area elements of corresponding resolution cells, then S can be approximated:

$$S \simeq \sum_{i=1}^{q} \sigma_i \tag{3.9}$$

Now the information provided by each of the spatially coded messages can be approximated:

$$I(S) = \log_2 N(S) \qquad \text{bits/message} \tag{3.10}$$

which is a function of S.

The rate of information per unit spatial element transmitted through the spatial channel is therefore

$$R(S) = \frac{I(S)}{S \Delta t} \qquad \text{bits/unit area/sec} \tag{3.11}$$

It is interesting to note that the rate changes as S varies. But it can be shown, for all practical cases, that the rate approaches a certain limit as S becomes sufficiently large. It is this limit, originally used by Shannon [1–3], that we adopt to define the capacity of an optical spatial channel:

$$C \triangleq \lim_{S \to \infty} \frac{\log_2 N(S)}{S \Delta t} \qquad \text{bits/unit area/sec} \tag{3.12}$$

This is the capacity that corresponds to the maximum average rate of information the spatial channel is capable of conveying under the constraints of Δt of Eq. (3.4). The spatial channel capacity in bits per unit spatial area can also be written

$$C \triangleq \lim_{S \to \infty} \frac{\log_2 N(S)}{S} \qquad \text{bits/unit area} \tag{3.13}$$

This is the limit we adopted for the spatial channel capacity as defined by Shannon [1–3].

It is noted that, if the optical channel is reduced to a single resolution cell (e.g., a single optical fiber), then the capacity of the channel can be calculated in a similar manner. For example, let us define a set of quantized signal levels E_i (irradiance levels) of duration t_i, where $i = 1, 2, \ldots, k$, and $t_i \geq \Delta t$ of Eq. (3.4). It is noted that we do not impose any restriction on the time duration; in other words, the duration may be identical. If we assume that the levels E_i are statistically independent over the time variable, then an arbitrary succession of signal levels can represent a coded message in the time domain. Thus for a long duration of T (i.e., where T is the sum of some t_i), a timely coded message transmitted through a channel element is simply the combination of all possible E_i's. Now let $N(T)$ be a total possible distinct combination of E_i's over T. If all the $N(T)$ possible combinations are assumed, *a priori*, to be equiprobable, then the information provided by each of the timely coded messages (i.e., the combination of E_i's) over T is

$$I(T) = \log_2 N(T) \quad \text{bits/message} \tag{3.14}$$

The corresponding rate at which information is transmitted through a single channel element is

$$R(T) = \frac{I(T)}{T} \quad \text{bits/sec} \tag{3.15}$$

It is clear that the rate varies with T. However, in all practical cases, the rate has a certain limit as T becomes increasingly large. This is the limit, as defined by Shannon [1–3], we used to define the channel capacity of a single resolution cell:

$$C = \lim_{T \to \infty} \frac{\log_2 N(T)}{T} \quad \text{bits/sec} \tag{3.16}$$

This is the capacity that corresponds to the maximum average rate of information it is possible to convey through a single-element optical channel.

It is also noted that, if we have $m \times n$ independent arrays of optical channels whose capacities are defined by Eq. (3.16), then the overall channel capacity is

$$C = \sum_{i=1}^{mn} C_i \quad \text{bits/sec} \tag{3.17}$$

It is also clear that, if all these $m \times n$ channels have an identical time duration of Δt over E_i ($t_1 = t_2 = \cdots = t_R = \Delta t$), then Eq. (3.16) can be reduced to Eq. (3.8), which is identical to the optical spatial channel. Since the time durations are all identical to the lowest limit Δt of the channel, the rate of information is maximum under the constraints of Eq. (3.4). This maximum rate of information transmission is the rate we have obtained for an optical channel with $m \times n$ arrays. It is noted that the result of Eq. (3.8) is also the capacity of the optical spatial channel under the constraints of Δt of Eq. (3.4).

We discuss the definition of channel capacity more thoroughly in Sec. 3.2, and conditions for its existence are illustrated.

3.2 OPTICAL MESSAGE IN SPATIAL CODING

Let us assume that we have made a selection of discrete signal levels (irradiances) and we want to apply these levels in synthesizing a spatially coded message. This is a typical example in optical spatial encoding. For instance, the distinct irradiance levels may correspond to a basic alphabet (e.g., the English alphabet), and their corresponding possible combinations are coded words. The irradiance levels could also represent basic digits whose sequences give rise to numbers.

It is noted that each of the signal levels used in information transmission represents a certain amount of energy (the cost) required for the transmission. The minimum amount of energy required for efficient information transmission through an optical channel depends on the physical nature of the channel. We investigate only the cost corresponding to the spatial elements of the optical signal. The basic problem in optical spatial communication is how to transmit as many messages as possible over a certain spatial domain S, and keep the required cost at a minimum.

Note that each signal level E_i occupies a set of spatial elements over S and that many irradiance levels may have spatial cells of identical size. Let us also assume that no constraints or restrictions are imposed on the possible distribution of the signal levels across the spatial domain (i.e., the levels are statistically independent). We now separate the signal levels according to the size of the spatial elements. Let h_i denote the number of signal levels that have spatial elements σ_i of the same size, where we assume $\sigma_1 < \sigma_2 < \cdots < \sigma_q$, σ_q representing the largest possible spatial element. Let M denote the total number of h_i:

$$M = \sum_{i=1}^{q} h_i \qquad (3.18)$$

We now seek the total number $N(S)$ of distinct spatial combinations of signal levels over the entire spatial domain S, and obtain

$$N(S) = \sum_{i=1}^{q} h_i N(S - \sigma_i) \qquad (3.19)$$

The above equation implies that the spatially coded messages corresponding to the spatial elements $S - \sigma_1$ can be completed by any one of h_1 signal levels to construct a message S, and in a similar manner for $h_2, h_3, \ldots, h_q$.

For simplicity, we assume that the σ_i's are integral values. It can be seen that Eq. (3.19) is a linear difference equation with constant coefficients of h_i. Thus it has σ_q independent solutions of the following type [4] (see Appendix A):

$$N_n(S) = Z_n^S \qquad n = 1, 2, \ldots, \sigma_q \tag{3.20}$$

and the general solution is the linear combination of Eq. (3.20):

$$N(S) = \sum_{n=1}^{q} K_n Z_n^S \tag{3.21}$$

where the K_n's are arbitrary constant coefficients.

We can now calculate the capacity of an optical spatial channel, which was defined in Sec. 3.1,

$$C = \lim_{S \to \infty} \frac{\log_2 N(S)}{S} \qquad \text{bits/unit area} \tag{3.22}$$

Let us consider an optical spatial channel in the case of spatial encoding. For a sufficiently large spatial domain S, the solution of Eq. (3.21) reduces to a dominant term:

$$N(S) = \lim_{S \to \infty} \sum_{n=1}^{q} K_n Z_n^S = K_1 Z_1^S \tag{3.23}$$

Alternatively, Eq. (3.23) can be written

$$N(S) = K_1 2^{\beta S} \qquad \text{as } S \to \infty \tag{3.24}$$

where $\beta = \log_2 Z_1$. By substituting Eq. (3.24) into Eq. (3.22), the capacity of the optical spatial channel is found:

$$C = \lim_{S \to \infty} \left(\frac{\log_2 K_1}{S} + \beta \right) = \beta \qquad \text{bits/unit area} \tag{3.25}$$

We noted that, if the optical spatial channel has a capacity C, and it is used to convey a spatial signal of I bits, then the best possible spatial encoding system that will enable us to utilize the channel efficiently has a spatial rate of $r = C/I$ signal levels per unit area.

In order to prove this general result, we discuss more precisely the physical aspects of a fundamental theorem regarding a noiseless spatial channel. We specify the basic properties of the encoding system so that the spatial channel will be more efficiently utilized. In the result, we show that the best spatial encoding techniques are the ones yielding the most probable distribution of signal levels. Moreover, certain basic encoding techniques are required for obtaining these optimum coding processes.

Let k be the total number of distinguishable signal levels for the spatial optical communication channel, and $n_1, n_2, \ldots, n_k$ the numbers of the respective signal levels $E_1, E_2, \ldots, E_k$ used in the spatially encoded messages. Then the total number of the coded messages is the total number of possible different combinations of the n_i's:

$$N = \frac{\left(\sum_{i=1}^{k} n_i\right)!}{n_1! n_2! \cdots n_k!} \tag{3.26}$$

where $\Sigma_{i=1}^{k} n_i$ is the total number of signal levels used in the encoding over S, which is equivalently equal to the number of resolution cells.

Now if we assume that the signal levels across the spatial domain are statistically independent, then their probability distribution is

$$P_j = \frac{n_j}{\sum_{i=1}^{k} n_i} \quad j = 1, 2, \ldots, k \tag{3.27}$$

with

$$\sum_{j=1}^{k} P_j = 1 \tag{3.28}$$

By means of Stirling's approximation [5], Eq. (3.25) can be approximated by

$$\log_2 N \approx -\left(\sum_{i=1}^{k} n_i\right) \sum_{j=1}^{k} P_j \log_2 P_j \quad \text{bits/message} \tag{3.29}$$

Since Eq. (3.29) is used to define the average amount of information provided by any one of the N possible spatially encoded messages, the average amount of information provided per signal is

$$I = \frac{\log_2 N}{\sum_{i=1}^{k} n_i} = -\sum_{j=1}^{k} P_j \log_2 P_j \quad \text{bits/signal level} \tag{3.30}$$

We can now present a simple example. We first consider that all the resolution cells over the spatial domain are equal in size, but that there are no restrictions on the signal levels. After this simple example, we treat a more general problem, namely, unequal resolution cells.

Let us now consider the case of resolution cells of equal size, $\sigma_1 = \sigma_2 = \cdots = \sigma$. In a large spatial domain S the spatially encoded messages contain a total number of S/σ resolution cells. In each of the resolution cells it is possible to encode any of the 1 to k signal levels. Thus the total number of possible spatially encoded messages (i.e., the total possible combinations) is

$$N(S) = k^{S/\sigma} \tag{3.31}$$

Then the corresponding capacity of the optical spatial channel of resolution cells of equal size is

$$C = \lim_{S \to \infty} \frac{\log_2 N(S)}{S} = \frac{\log_2 k}{\sigma} \quad \text{bits/unit area} \tag{3.32}$$

It is noted that, in defining the spatial channel capacity, *a priori* probability distributions of the signal levels across the spatial domain are assumed equiprobable. It is clear that under this equiprobability assumption the average amount of information provided by the spatial information source is maximum:

$$I_{\max} = -\sum_{i=1}^{k} P_i \log_2 P_i = \log_2 k \quad \text{bits/signal level} \tag{3.33}$$

Since each signal level is assumed to encode in a resolution cell of the same size, the rate at which information is provided by the source is

$$R = \frac{I_{\max}}{\sigma} = \frac{\log_2 k}{\sigma} \quad \text{bits/unit area} \tag{3.34}$$

It can be seen that Eq. (3.34) essentially results from the optical spatial channel capacity of Eq. (3.32). It is noted that this result basically comes from the apparent optimization of the spatial information source, in which the signal levels are taken to be equal in probability. Furthermore, the rate of transmission is

$$r = \frac{C}{I_{\max}} = \frac{1}{\sigma} \quad \text{signal levels/unit area} \tag{3.35}$$

The basic significance of Eq. (3.35) is that we simply cannot transmit more than r signal levels per unit area without introducing significant error.

It is also noted that, for a finite spatial domain S, the spatial channel capacity can be written

$$C(S) = \frac{S}{\sigma} \log_2 k \quad \text{bits} \tag{3.36}$$

Equation (3.36) indicates the maximum allowable rate at which spatial information can be transmitted through the channel without causing significant error.

With this simple illustration, it can be seen that the best spatial encoding scheme involves application of the equiprobability of the signal levels at the input end of the channel. This is a significant result, based on the basic information theory of Shannon [1–3], for efficient utilization of the channel.

In concluding this example, we note that the equiprobability distribution of the signal levels across the spatial domain yields the maximum spatial information. This maximum spatial information in fact is the asymptotic result of obtaining the spatial channel capacity of Eq. (3.32). It is noted that in the course of the calculation we did not impose an *a priori* probability distribution of the signal levels. But it should be cautioned that the asymptotic method of determining the capacity of the channel is an *average notion*, which is applied to a large spatial domain. This asymptotic approximation is the basic feature in calculation of the spatial channel capacity. We show in the following, a more general case for resolution cells of different sizes, that the spatial averaging notion is still applicable in the calculation of channel capacity.

3.3 SPATIAL CHANNEL WITH RESOLUTION CELLS OF DIFFERENT SIZES

For the more general case of resolution cells of different sizes, the illustration is more involved. Therefore we treat this example in a separate section.

We discuss this case in detail and show that the result obtained in Sec. 3.2 still applies. The most efficient spatial encoding is the one that yields the most probable *a priori* probability distribution for the different signal levels across the spatial domain.

Let us now assume an encoded message that occupies the spatial domain S of the channel. Each signal level (irradiance) E_i, where $i = 1, 2, \ldots, k$, has resolution cells of a certain size and many signal levels have cells of the same size. Let n_j, where $j = 1, 2, \ldots, k$, be the number of signal levels having

Optical Spatial Channel and Encoding Principles

resolution cell of the same size σ_j. The total spatial domain is then

$$S = \sum_{j=1}^{k} n_j \sigma_j \qquad (3.37)$$

where $\sigma_1 < \sigma_2 < \cdots < \sigma_k$. The total number of signal levels across the spatial domain S is

$$M = \sum_{j=1}^{k} n_j \qquad (3.38)$$

The probability distribution of the signal levels over the spatial domain S is therefore

$$P_j = \frac{n_j}{M} \qquad (3.39)$$

with

$$\sum_{j=1}^{k} p_j = 1 \qquad (3.40)$$

By substituting $n_j = MP_j$ in Eq. (3.37) we have

$$S = M \sum_{j=1}^{k} \sigma_j P_j \qquad (3.41)$$

Since the total number of different spatially coded messages (i.e., combinations of n_j over M) is

$$N = \frac{M!}{n_1! n_2! \cdots n_k!} \qquad (3.42)$$

the logarithm of N, by Stirling's approximation [5], can be approximated by

$$\log_2 N = -M \sum_{j=1}^{k} P_j \log_2 P_j \qquad (3.43)$$

If we assume that the spatially encoded messages over the spatial domain are statistically independent, the average amount of information provided by one of the N encoded messages is

$$I = \log_2 N \quad \text{bits/message} \qquad (3.44)$$

It is noted that all the spatially encoded N messages occupy the same spatial domain S, since they are assumed to have fixed numbers of n_j of size σ_j over

S, where $j = 1, 2, \ldots, k$, and they are assumed, *a priori*, to be equiprobable across the spatial domain of S.

Now in the search for the most probable distribution of the input signal ensemble for the maximum amount of information per encoded message, we consider the following variational technique [6]:

$$d(\log_2 N) - \alpha d\left(\sum_{j=1}^{k} P_j\right) - \beta\, d(S) = 0 \tag{3.45}$$

where the parameters α and β are known as Lagrange multipliers. We note that $M, p_1, p_2, \ldots, p_k$ are the variables for consideration. Thus, by differentiating Eq. (3.43) with respect to these variables, we have

$$d(\log_2 N) = -dM \sum_{j=1}^{k} P_j \log_2 P_j - M \sum_{j=1}^{k} (\log_2 e + \log_2 P_j)\, dP_j \tag{3.46}$$

By substituting Eq. (3.46) into Eq. (3.45) we have

$$-dM\left[\sum_{j=1}^{k} P_j(\log_2 P_j + \beta\sigma_j)\right]$$
$$+ \sum_{j=1}^{k} dP_j[-\alpha - \beta M\sigma_j - M(\log_2 e + \log_2 P_j)] = 0$$
$$\tag{3.47}$$

With regard to the independent variables dM and dP_j it can be seen that

$$\sum_{j=1}^{k} P_j(\log_2 P_j + \beta\sigma_j) = 0 \tag{3.48}$$

and

$$\alpha + \beta M\sigma_j + M(\log_2 e + \log_2 P_j) = 0 \tag{3.49}$$

Equation (3.49) can be written

$$-\log_2 P_j = \log_2 e + \beta\sigma_j + \frac{\alpha}{M} \tag{3.50}$$

By substituting Eq. (3.50) into Eq. (3.48) we have

$$\left(\log_2 e + \frac{\alpha}{M}\right)\sum_{j=1}^{k} P_j = 0 \tag{3.51}$$

But since Eq. (3.40), we conclude that

$$\alpha = -M\log_2 e \tag{3.52}$$

Now Eq. (3.50) can be reduced to
$$-\log_2 P_j = \beta \sigma_j \tag{3.53}$$
P_j can be written
$$P_j = 2^{-\beta \sigma_j} \tag{3.54}$$
and β can be determined by
$$\sum_{j=1}^{k} P_j = \sum_{j=1}^{k} 2^{-\beta \sigma_j} = 1 \tag{3.55}$$
Let
$$Z = 2^\beta \tag{3.56}$$
Then Eq. (3.55) can be written
$$\sum_{j=1}^{k} Z^{-\sigma_j} = 1 \tag{3.57}$$
which is essentially the same as Eq. (3.21), with all $K_n = 1$.

If the spatial encoding is over a very large spatial domain, then the input signal ensemble will approach asymptotically the most probable distribution at which a maximum amount of spatial information will be provided by the encoded messages. Substitution of Eq. (3.54) into Eq. (3.43) yields
$$\log_2 N = M \sum_{j=1}^{k} \beta \sigma_j 2^{-\beta \sigma_j} - M\beta \sum_{j=1}^{k} \sigma_j P_j \tag{3.58}$$
By definition, the average size of the resolution cells is
$$\bar{\sigma} = \sum_{j=1}^{k} \sigma_j P_j \tag{3.59}$$
and
$$S = \bar{\sigma} M \tag{3.60}$$
Then Eq. (3.58) can be written
$$\log_2 N = \beta S \tag{3.61}$$
which is equivalent to
$$N = 2^{\beta S} = Z_1^S \tag{3.62}$$
where $\beta = \log_2 Z_1$.

It is noted that, with the most probable distribution of the input signal ensemble, N is essentially the same as $N(S)$ of Eq. (3.24). Therefore the spatial channel capacity is

$$C = \lim_{S \to \infty} \frac{\log_2 N}{S} = \beta \quad \text{bits/spatial area} \tag{3.63}$$

From this result, it can be seen that, for spatial encoding over large domains, the coded message is probably used more frequently than any other encoding. Therefore all the other possible distributions can simply be disregarded. The best coding is the one that achieves the most efficient distribution of signal levels:

$$\frac{\log_2 P_j}{\sigma_j} = -\beta \tag{3.64}$$

where β is a constant.

From Eq. (3.55), we learn that the largest real root of β corresponds to the largest value of Z. In view of Eq. (3.61), the largest root of β gives rise to the largest value of N. Under these conditions, it yields the most probable distribution of the signal levels.

Now if we use the most efficient coding and the most probable distribution of the signal levels, then the amount of information per signal level provided by this spatial encoding over S will be

$$I = \sum_{j=1}^{k} P_j \log_2 P_j = \beta \bar{\sigma} \quad \text{bits/signal level} \tag{3.65}$$

where $\bar{\sigma}$ is the average size of the resolution cells, as defined by Eq. (3.59). The corresponding rate of spatial information in a signal level per unit area is

$$r = \frac{C}{I} = \frac{1}{\bar{\sigma}} \tag{3.66}$$

which is essentially the same as the rate given by Eq. (3.35).

In concluding this section we point out that the coincidence of the average and most probable distribution of the signal levels is a well-known result based on the law of large numbers in statistics [7]. This is also a basic and important concept of entropy theory in statistical thermodynamics [8].

3.4 MATCHING A CODE WITH A SPATIAL CHANNEL

In Sec. 3.3 we arrived at the conclusion that, for the most efficient spatial encoding, the most probable distribution of the input signal ensemble over the spatial domain is that of Eq. (3.54) or, equivalently, Eq. (3.64). This spatial encoding is just as important as the problem of matching the refractive index of an optical component. However, the result we obtained was rather general. In principle it can be applied to all coding problems. In order to demonstrate the applicability of the result to a general spatial channel, we discuss a few simple examples.

Let us recall the efficient encoding conditions:

$$P_j = 2^{-\beta \sigma_j} \qquad j = 1, 2, \ldots, k \tag{3.67}$$

where P_j can be considered, *a priori*, the probability of the spatial signal, σ_j is the corresponding spatial cell, and β is an arbitrary constant.

If the most probable input spatial signal has a probability of P_j, then $-\log_2 P_j$ will be relatively small. In the most efficient spatial encoding, the corresponding σ_j should also be made smaller. However, for the most improbable one, σ should be made larger.

In view of the probability distribution of the input spatial signals, we can obtain in principle a corresponding variation in spatial encoded messages, for example, in a code of k levels. This would apply to a spatial channel capable of transmitting k quantized signal levels, namely, E_i, $i = 1, 2, \ldots, k$, irradiance levels.

Since our spatial encoding may involve resolution cells of different sizes, it is customary to provide some means for identifying any one of the input spatial signals. For simplicity, we assume a rectangular spatial domain with square array of resolution cells of the same size. It is also assumed that k levels of encoding take place on a row-by-row basis by sequentially scanning across the spatial domain. In this case a specified signature to characterize the end of each code word is added. Therefore, in a long encoded message, the average size of the resolution cell per input signal can be obtained, such as

$$\bar{\sigma} = \sum_{j=1}^{k} \sigma_j P_j \tag{3.68}$$

In order to illustrate this basic technique we present two simple examples.

We first consider the encoding of a set of letters in the English alphabet sequentially in a rectangular binary spatial domain ($k = 2$). In this encoding process, we use a spatial channel capable of transmitting only one type of

Table 3.1 Probability of occurrence of English Letters

Letter	Probability	Code	Letter	Probability	Code
Space	0.1859	111			
A	0.0642	01111	N	0.0574	01011
B	0.0127	0001111	O	0.0632	0011
C	0.0218	000111	P	0.0152	0100111
D	0.0317	001111	Q	0.0008	01010011
E	0.1031	011	R	0.0484	00011
F	0.0208	000011	S	0.0514	010111
G	0.0152	0010011	T	0.0796	0111
H	0.0467	010011	U	0.0228	0101111
I	0.0575	00111	V	0.0083	0001011
J	0.0008	01010111	W	0.0175	0010111
K	0.0049	0000111	X	0.0013	000011
L	0.0321	001011	Y	010164	0100011
M	0.0198	0101011	Z	0.0005	01001111

Source: Reza [9].

spatial irradiance in each of the resolution cells; with no irradiance the cell is regarded as a 0, and with irradiance it is regarded as a 1.

Since the spatial coding corresponds to different lengths for each input signal, we must provide some method for recognizing signal terminations. For example, a double, triple, or quadruple of 1 may be used for the termination of each spatial encoded letter. It is also noted that a single 1 is regarded as a signal proper but not as a signature for termination. We now consider the *a priori* probability distribution of a set of English letters derived from a book, as shown in Table 3.1 [9]. The corresponding binary codes are also tabulated. It can be seen from this table that the spatial coding varies from three to eight cells per letter. In a long message sequentially encoded on a spatial rectangular domain, the average length per letter is

$$\bar{\sigma} = \sum_{j=1}^{27} \sigma_j P_j \simeq 4.65 \text{ cells/letter} \tag{3.69}$$

Since we are using binary coding, $\bar{\sigma}$ is also the average information provided by the encoded letter:

$$I' = \bar{\sigma} \log_2 2 = 4.65 \text{ bits/letter} \tag{3.70}$$

Optical Spatial Channel and Encoding Principles

Table 3.2 A Possible Ternary Code

Letters	Code	Letters	Code
Space	20		
A	210	N	220
B	12210	O	120
C	2210	P	11210
D	2110	Q	21220
E	10	R	1120
F	2220	S	1210
G	12120	T	110
H	1220	U	11110
I	1110	V	12220
J	21210	W	12110
K	21110	X	21120
L	2120	Y	11220
M	11120	Z	22110

It is also noted that the average amount of information provided by the letters is

$$I = \sum_{j=1}^{27} P_j \log_2 P_j = 4.03 \text{ bits/letter} \quad (3.71)$$

It can be seen from Eq. (3.70) that the entropy of our binary coding is still larger than the entropy of the source of Eq. (3.71). If we took the binary coding process to be equal in word length, then it would require five cells for each letter. In this case we can see that the coding is even worse than the one chosen at the beginning. Nevertheless, there exists an optimum technique so that the encoding may approach the theoretical limit of Eq. (3.71). Obviously, in our example the coding is by no means optimum.

In the second example, we adopt the same probability distribution of the English alphabet as in the previous example, but we use a ternary code ($k = 3$). In this encoding process, we employ a spatial channel transmitting three levels of irradiance, namely, E_0, E_1, and E_2. Let E_0 correspond to a 0, E_1 to a 1, and E_2 to a 2. The ternary codes are listed in Table 3.2. We used E_0 (i.e., zero) as a termination signature for the encoded letters. It is noted that without this signature we are left with a binary coding for the letters. The average length of the encoding is

$$\bar{\sigma} = \sum_{j=1}^{27} \sigma_j P_j \simeq 3.3 \text{ cells/letter} \quad (3.72)$$

The average information provided by this encoding procedure is

$$I' = \bar{\sigma} \log_2 3 = 5.25 \text{ bits/letter} \tag{3.73}$$

which is higher than in the previous example. Therefore it may be seen that for multisignal level encodings, it does not necessarily reflect the merit of coding. It is the optimum encoding process that reflects the merit.

It is true for most sequential encoding techniques that, for different lengths, a special signature is required for the termination of each encoded signal. This is the price we pay for lengthening the encoding scheme. For example, in our case, zero (i.e., E_0) appears less frequently in the encoded signals. It is for this reason that the average information provided by the encoded letters is greater than that of the previous example. Nevertheless, a more elaborate encoding procedure may improve the result.

In concluding this section we also note that the *coding efficiency* can be defined:

$$\eta = \frac{I}{I'} \tag{3.74}$$

Where I and I' are the average information provided by the signals and encoded signals, respectively. It is also possible to define the redundancy of the encoding:

$$\text{Redundancy} = 1 - \eta \tag{3.75}$$

Therefore in the previous examples the coding efficiencies for the binary and ternary cases are

$$\eta = \frac{4.03}{4.65} = 0.867 \quad \text{or } 86.7\% \tag{3.76}$$

and

$$\eta = \frac{4.03}{5.25} = 0.768 \quad \text{or } 76.8\% \tag{3.77}$$

The corresponding redundancies are

$$\text{Redundancy} = (1 - 0.867)\% = 13.3\% \tag{3.78}$$

and

$$\text{Redundancy} = (1 - 0.768)\% = 23.2\% \tag{3.79}$$

Since optimum encoding processes are beyond the scope of this book, we refer the reader to the excellent text by Peterson [10].

REFERENCES

1. C. E. Shannon, "A Mathematical Theory of Communication," *Bell Syst. Tech. J.*, vol. 27, 379–423, 623–656 (1948).
2. C. E. Shannon, "Communication in the Presence of Noise," *Proc. IRE*, vol. 37, 10 (1949).
3. C. E. Shannon and W. Weaver, *The Mathematical Theory of Communication*, University of Illinois Press, Urbana, 1949.
4. C. H. Richardson, *An Introduction to the Calculus of Finite Differences*, Van Nostrand, Princeton, N. J., 1954.
5. W. Kaplan, *Advanced Calculus*, Addison-Wesley, Reading, Mass., 1952.
6. N. I. Akhiezer, *The Calculus of Variations*, Blaisdell, New York, 1962.
7. E. Parzen, *Modern Probability Theory and Its Applications*, John Wiley, New York, 1960.
8. F. W. Sears, *Thermodynamics, the Kinetic Theory of Gases, and Statistical Mechanics*, Addison-Wesley, Reading, Mass., 1953.
9. F. M. Reza, *An Introduction to Information Theory*, McGraw-Hill, New York, 1961.
10. W. W. Peterson, *Error-Correcting Codes*, MIT Press, Cambridge, Mass., 1961.
11. L. Brillouin, *Science and Information Theory*, Academic, New York, 1956.
12. D. Gabor, "Light and Information," in E. Wolf, ed., *Progress in Optics*, vol. I, North-Holland, Amsterdam, 1961.

4
Entropy and Information

In the preceding chapters we discussed the basic definition of information, spatial channel capacity, and the principles of coding. The fundamental constraints of these definitions were also discussed. In this chapter we deal with the relationship between information theory and physical science.

The measure of information, as Shannon [1–3] defined it, has an intimate relationship with entropy theory in statistical thermodynamics. Therefore information theory and thermodynamics must have some common points of interest. In order to discuss the basic aspects of entropy as related to information, we first review the two fundamental laws or principles of thermodynamics [4, 5].

4.1 FUNDAMENTAL LAWS OF THERMODYNAMICS

We state, without proof, the two fundamental laws of thermodynamics. The *first law* of thermodynamics may be regarded as the law of *energy conservation*. In very general terms, the first law asserts that the net flow of energy across the boundary of a system is equal to the change in energy of the system. For this reason, in thermodynamics it is sufficient to consider only two types of energy flow across the boundary, namely, the work done on or by the system, and the heat flow supplied by radiation or conduction to or by the system.

The *second law* of thermodynamics can be regarded as Carnot's principle [4, 5], which asserts that, in every process that takes place in an *isolated* system, the entropy, as defined by Clausius [4, 5] of the system either increases or remains unchanged. In other words, the entropy of an isolated system never decreases.

According to Kelvin [4, 5] the second law is essentially the law of *energy degradation*. That is, although the total energy in a closed isolated system remains constant, energy degradation for all *irreversible* processes

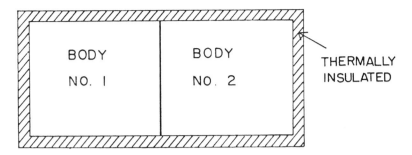

Fig. 4.1 An irreversible thermodynamic process.

results. We say the energy is degraded because it tends to a common level, hence cannot perform any work by flowing from a higher to a lower (temperature) level.

Furthermore, if an isolated system is in such a state that its entropy is at a maximum, then a change from that maximum entropy state to a lower entropy state is *not* possible, since the system is closed and isolated.

Now if a system receives an amount of heat ΔQ (by conduction, radiation, or any physical means), then the increase in entropy in the system can be written [4, 5]

$$\Delta S = \frac{\Delta Q}{T} \qquad (4.1)$$

where $T = C + 273$ is the absolute temperature in Kelvins, and C is the temperature in degrees Celsius.

The following examples will clarify the reader's definition of entropy. Let us consider two bodies that are in contact, but as a whole are isolated, as shown in Fig. 4.1. These two bodies may be assumed to conduct heat exchange and work done between one another. For convenience, we denote these two bodies 1 and 2. Then by the first law (energy conservation) of thermodynamics, we conclude that

$$\Delta Q_1 - \Delta W_1 + \Delta Q_2 - \Delta W_2 = 0 \qquad (4.2)$$

where ΔQ and ΔW are the corresponding heat input and work done by the bodies, and the subscripts 1 and 2 denote the respective bodies. It is noted from Eq. (4.2) that there is no change in the overall energy. However, the changes in entropy within the two bodies are

$$\Delta S_1 = \frac{\Delta Q_1}{T_1} \qquad (4.3)$$

and
$$\Delta S_2 = \frac{\Delta Q_2}{T_2} \tag{4.4}$$

Since the two bodies as a whole are isolated, by the second law of thermodynamics we have

$$\Delta S = \Delta S_1 + \Delta S_2 > 0 \tag{4.5}$$

Thus the entropy of the whole system increases.

By a slight change in the above example, we let the temperature of body 1 be higher than that of body 2, $T_1 > T_2$. Then the heat flow is conducted from body 1 to body 2. It is assumed that no work is performed in either of these bodies, $\Delta W_1 = \Delta W_2 = 0$, so the heat transfer from body 1 to body 2 is therefore

$$\Delta Q_2 - \Delta Q_1 = 0 \tag{4.6}$$

where ΔQ_2 is the amount of heat *input* to body 2, a positive quantity, and $-\Delta Q_1$ is the amount of heat *output* from the body 1, a negative quantity. Thus the corresponding entropy changes of these bodies are

$$\Delta S_1 = -\frac{\Delta Q_1}{T_1} = -\frac{\Delta Q_2}{T_1} \tag{4.7}$$

and

$$\Delta S_2 = -\frac{\Delta Q_2}{T_2} = -\frac{\Delta Q_1}{T_2} \tag{4.8}$$

Although the entropy of body 1 decreases, we can see that the overall entropy, that is, the combined entropy of these two bodies, increases.

Since by Eq. (4.6), $\Delta Q_2 = \Delta Q_1$,

$$\Delta S = \Delta S_1 + \Delta S_2 = \Delta Q_1 \left(\frac{1}{T_2} + \frac{1}{T_1} \right) > 0 \tag{4.9}$$

where $T_1 > T_2$. This is an example of an irreversible process of an isolated system in the thermodynamic sense, for heat cannot flow from a lower to a higher temperature.

With respect to the same example, if we let the temperature of both bodies remain constant, $T_1 = T_2 = T$, and we assume that there is heat input but no work done in body 1, $\Delta W_1 = 0$, and $\Delta Q_1 > 0$, and there is work done but no heat exchange in body 2, $\Delta W_2 > 0$, and $\Delta Q_2 = 0$, then by the first law of thermodynamics we see that

$$\Delta Q_1 - \Delta W_2 = 0 \tag{4.10}$$

In this case, the entropy changes within these two bodies are

$$\Delta S_1 = \frac{\Delta Q_1}{T} = \frac{\Delta W_2}{T} \tag{4.11}$$

and

$$\Delta S_2 = 0 \tag{4.12}$$

Once again, as a whole, the entropy of the isolated system increases:

$$\Delta S = \Delta S_1 + \Delta S_2 = \frac{\Delta Q_1}{T} = \frac{\Delta W_2}{T} > 0 \tag{4.13}$$

So one sees that the exchange of heat and work done is an irreversible process which results in an entropy increase in an isolated system.

It is noted that the preceding illustrations are typical in thermodynamics. We have shown that it is impossible for the entropy to decrease in an isolated system. The second example shows that the entropy within one of the bodies essentially decreases (i.e., the problem of heat flow but no work done), $\Delta S_1 < 0$, but the overall entropy increases, $\Delta S = \Delta S_1 + \Delta S_2 > 0$.

The increase in entropy has been regarded as the degradation of energy by Kelvin. It is noted that the word *entropy* was originally introduced by Clausius, and he may have intended that it be used for a negative interpretation of the entropy concept. Therefore it is occasionally found to be much easier to discuss the problem from the point of view of negative entropy, or simply *negentropy* ($N \triangleq -S$). Historically, the importance of the negentropy concept was first recognized by Tait [6], a close associate of Kelvin. Negentropy represents the quality or grade of energy of an isolated system and must always decrease or remain constant. In other words, negentropy may be regarded as the opposite interpretation of the second law of thermodynamics.

From the above examples, it is noted that, for an isolated system (i.e., the combination of the two bodies in our example) to perform work (e.g., electrical, mechanical) or even thermal dissipation, it must contain a source of negentropy. This source of negentropy is used to do electrical or mechanical work or is simply lost by thermal dissipation. An isolated system containing a charged battery and a dc motor with a load is a good example. We can consider the charged battery a source of negentropy (i.e., the entropy in the battery must increase).

The negentropy principle of information was first emphasized by Brillouin [7] and is fully discussed in his excellent book *Science and Information Theory* [8]. Historically, the negative entropy concept may be due to Schrödinger [9] and his interesting book *What Is Life?* He pointed

Entropy and Information 99

out that, for a living organism to sustain life, it needs to draw negative entropy from its environment continually. However, in our version, if a living organism is searching for food, then it requires a certain amount of negentropy from some source for it to reach for the food. The amount of negentropy required makes up for the work it has performed in searching for the food. Since energy as a whole is conserved, the energy provided by the food is not what enables the organism to reach for it—the negentropy required is important.

4.2 PHYSICAL ENTROPY AND INFORMATION

The basic definition of information was given in earlier chapters a strictly statistical point of view. We considered an example of N possible symbols (or outcomes) of an information source (or experiment). If these N possible symbols are assumed, *a priori*, to be equiprobable, then the average amount of information provided by the source will be

$$I_0 = -\sum_{j=1}^{N} P_j \log_2 P_j = \log_2 N \quad \text{bits/symbol} \tag{4.14}$$

where $P_j = 1/N$.

A certain amount of information I is acquired with which it may be possible to reduce the N symbols to a smaller set of symbols M. If it is also assumed, *a priori*, that the M symbols are equiprobable, then the average amount of information provided by the M symbols will be

$$I_1 = \sum_{j=1}^{M} P_j \log_2 P_j = \log_2 M \quad \text{bits/symbol} \tag{4.15}$$

where $P_j = 1/M$, and $N > M$. The amount of information required for this information reduction is

$$I = I_0 - I_1 = \log_2 \frac{N}{M} \quad \text{bits/symbol} \tag{4.16}$$

We are seeking the relationship between information and entropy. For this reason we consider mainly only information connected with certain physical problems in which entropy theory can be easily treated. In order to derive the relationship between information and entropy, we consider a physical system in which equiprobability in complexity of the structures has been established *a priori*. It is noted in the previous example (where the N possible outcomes are reduced to M) that the system *cannot* be isolated, since the second law of thermodynamics prevails. The corresponding entropies for the N and M equiprobable states are

$$S_0 = k \ln N \tag{4.17}$$

and

$$S_1 = k \ln M \tag{4.18}$$

where $N > M$, and k is Boltzmann's constant. Thus

$$S_0 > S_1 \tag{4.19}$$

It can be seen that the entropy decreases when information I is obtained. However, this amount of information I must be furnished by some external source whose entropy increases. The decrease in entropy in our system is obviously derived from the information I we have acquired from some external sources:

$$\Delta S = S_1 - S_0 = -kI \ln 2 \tag{4.20}$$

Or we write, equivalently,

$$S_1 = S_0 - kI \ln 2 \tag{4.21}$$

where $S_0 > S_1$.

However, the information I is required to be proportional to the decrease in entropy ΔS in our system. This is the basic connection between entropy and information theory. Equation (4.20) or, equivalently, Eq. (4.21) can be regarded as an expression of the *entropy theory of information*. We discuss this relationship in more detail as we apply Eq. (4.21) to some physical problems. We show that the information I and the change in entropy ΔS can simply be traded.

4.3 TRADING ENTROPY WITH INFORMATION

We have established an intimate relationship between information and entropy [Eq. (4.21)]. The amount of information acquired in reducing the entropy of a physical system must be derived from some external source.

By the second law of thermodynamics, if we isolate the entire system, which includes sources of I, then for any further evolution within the whole system the entropy will increase or remain constant:

$$\Delta S_1 = \Delta(S_0 - kI \ln 2) \geq 0 \tag{4.22}$$

From Eq. (4.22) we see that any further increase in entropy ΔS_1 can be due to ΔS_0 or ΔI or both. Although in principle it is possible to distinguish the changes in ΔS_0 and ΔI separately, in some cases the separation of the changes due to ΔS_0 and ΔI may be difficult to discern.

It is interesting to note that, if the initial entropy S_0 of the system corresponds to some complexity of a structure but not the maximum, and if S_0 remains constant ($\Delta S_0 = 0$), then after a certain free evolution without the influence of external sources, from Eq. (4.22), we will have

$$\Delta I \leq 0 \tag{4.23}$$

since $\Delta S_0 = 0$.

From Eq. (4.23), we see that the changes in information ΔI are negative, or decreasing. The interpretation is that, when we have no *a priori* knowledge of the system complexity, the entropy S_0 is assumed to be maximum (i.e., the equiprobable case). Hence the information provided by the system structure is maximum. Therefore $\Delta I \leq 0$ is due to the fact that, in order to increase the entropy of the system, $\Delta S_1 > 0$, a certain decrease in information is needed. In other words, information can be provided or transmitted (a source of negentropy) only by increasing the entropy of the system. However, if the initial entropy S_0 is at a maximum, then $\Delta I = 0$ and the system cannot be used as a source of negentropy.

In Sec. 4.4 we present several examples illustrating that entropy and information can be interchanged. Information can be obtained *only* by increasing the entropy of certain physical devices. This statement is expressed by the simple symbolic equation

$$\Delta I \rightleftharpoons \Delta S \tag{4.24}$$

We summarize the relationship between entropy and information theory as follows.

1. The entropy corresponds to the average amount of information provided, as defined by Shannon [1-3].
2. The increase in entropy in a physical system may be used as a source of negentropy to provide information, or vice versa.
3. Work done or energy can be shown to be related to information:
$$\Delta W = \Delta Q = T \Delta S = I k T \ln 2 \tag{4.25}$$
   where T is the thermal noise temperature. Thus, from the above equation, we see that with higher thermal noise T the amount of energy required for the transmission of information I is higher.

From Eq. (4.22), the equality holds only for a reversible process; otherwise the entropy increases.

4.4 TYPICAL EXAMPLES

In order to provide examples of the relationship between entropy and information, we first discuss a general information problem.

For simplicity, let us begin with a communication example, say an oral conversation. An individual possesses a certain piece of information that is to be shared with a friend. We illustrate with a step-by-step process how this piece of information may be transmitted.

First an appropriate coding (e.g., the right language) must be selected, and then this piece of information must be transformed, or encoded, into an appropriate speech signal suitable for the acoustical communication channel. However, in the signal transformation process certain errors may be introduced. These errors may represent certain information loss due to the encoding process. It is noted that the loss of information occurs before the information is transmitted through the channel.

Second, a certain loss of information may also occur as a result of noise disturbances in the channel, for example, acoustical noise, which depends on the nature of the communication channel.

Third, the receiver (the friend) may have difficulty in decoding the information. For example, because of a defect in the receiving system (the ears), the coding may be misinterpreted or even missed by the receiver. This represents additional information loss.

Fourth, after a period of time elapses, the receiver (the friend), may forget some of the information that has been received.

So it can be seen that there is a certain amount of information loss at each stage. Of course, the most efficient communication we are aiming at is one without any loss during information transmission.

This example is a very common one and applies to our daily communication. Another trivial example involves a broadcaster reporting daily news events on the radio. Not all the information conveyed by the broadcaster is equally valuable to the listener, so some of it may be ignored or overlooked. It is noted that our definition of information was not originally intended to include this value aspect, but it is possible to extend the theory in this way. However, the discussion of such a problem is beyond the scope of this book. Interested readers may refer to the existing literature.

We now discuss a few typical examples from the statistical thermodynamics point of view, to which the relationship between entropy and information can be easily applied.

Entropy and Information 103

Let us consider, for example, an ideal monatomic gas, assumed to exist in a thermally isolated container. Under steady-state conditions, the familiar entropy equation can be written [10, 11]

$$S = \frac{5Nk}{2} + Nk \ln\left[\frac{Vg}{N}\left(\frac{4\pi mE}{3Nh^2}\right)^{3/2}\right] \tag{4.26}$$

where N is the total number of atoms in the container, k is Boltzmann's constant, V is the volume of the container, m is the atomic mass, E is the total energy, h is Planck's constant, $g = 2j+1$ is the number of indistinguishable ground states of the atom, and j is the momentum moment.

The value of g is worth mentioning. If the ground state is not degenerate, then $j = 0$, and $g = 1$. In this case the atom has no momentum moment in the ground state. However, if $j \neq 0$, then there are $g = 2j+1$ ground states, and we assume *a priori* that these ground states are equiprobable.

We now discuss the first example. We assume that we had *a priori* knowledge of the state of the gas. For instance, we may know that at a precise moment the gas occupies a volume V_0, and that at another moment it occupied a larger volume V_1:

$$V_1 = V_0 + \Delta V \tag{4.27}$$

where ΔV is the enlarging portion.

By the entropy equation (4.26), the corresponding entropies can be written, respectively,

$$S_0 = Nk(\ln V_0 + K) \tag{4.28}$$

and

$$S_1 = Nk(\ln V_1 + K) \tag{4.29}$$

where

$$K = \tfrac{5}{2} + \ln\left[\frac{g}{N}\left(\frac{4\pi mE}{3Nh^2}\right)^{3/2}\right] \quad \text{and} \quad S_0 < S_1$$

The initial entropy S_0 is somewhat smaller than the final entropy S_1 after the gas expansion. The corresponding increase in entropy of this process is, from Eq. (4.20),

$$\Delta S = S_1 - S_0 = Nk(\ln V_1 - \ln V_0) > 0 \tag{4.30}$$

This increase in entropy may be considered the loss in information of the process:

$$I = \frac{\Delta S}{k \ln 2} \tag{4.31}$$

Evaluation of the above results is made when the steady state of the gas is reached, that is, after a uniform density distribution is established throughout the final volume V_1.

Again we see that an increase in entropy of a process corresponds to certain information loss. In this example, it may be said that, as the volume expanded, the gas progressively lost or forgot the information.

However, we can see that, if the volume progressively shrinks, the gas obtains additional information, since the entropy progressively decreases. But we have seen from Eq. (4.20) that this is possible only if a certain amount of information is acquired from some external source.

We use the same monatomic gas for the second example, and consider the problem of diffusion. For simplicity, we assume that two different types of gases are involved, but that they have an identical mass m and the same g factor. We let N_1 atoms of type 1 occupy volume V_1, and N_2 atoms of type 2 occupy volume V_2. Thus the total atoms and total volume are

$$N = N_1 + N_2 \tag{4.32}$$

and

$$V = V_1 + V_2 \tag{4.33}$$

Let us define the corresponding probabilities:

$$P_1 = \frac{N_1}{N} = \frac{V_1}{V} \tag{4.34}$$

and

$$P_2 = \frac{N_2}{N} = \frac{V_2}{V} \tag{4.35}$$

where $P_1 + P_2 = 1$.

From Eqs. (4.34) and (4.35) we have

$$\frac{V_1}{N_1} = \frac{V_2}{N_2} = \frac{V}{N} \tag{4.36}$$

Let us also assume that the law of *energy equipartition* is applied. Then we conclude:

$$\frac{E_1}{N_1} = \frac{E_2}{N_2} = \frac{E}{N} \tag{4.37}$$

Entropy and Information

where

$$E = E_1 + E_2$$

It is noted from Eqs. (4.36) and (4.37) that the two gases are under the same pressure and temperature. For simplicity, we let volume V be constant throughout the gas diffusion process. Thus the entropies before and after the diffusion process are

$$S_0 = Nk\left(\ln \frac{V}{N} + G\right) \tag{4.38}$$

and

$$S_1 = N_1 k\left(\ln \frac{V}{N_1} + G\right) + N_2 k\left(\ln \frac{V}{N_2} + G\right) \tag{4.39}$$

where

$$G = \tfrac{5}{2} + \ln\left[g\left(\frac{4\pi mE}{3Nh^2}\right)^{3/2}\right]$$

From Eqs. (4.36) and (4.38), Eq. (4.39) can be written

$$S_1 = S_0 + k\left(N_1 \ln \frac{N}{N_1} + N_2 \ln \frac{N}{N_2}\right) \tag{4.40}$$

By substituting Eqs. (4.34) and (4.35) into Eq. (4.40), we have

$$S_1 = S_0 - Nk(P_1 \ln P_1 + P_2 \ln P_2) \tag{4.41}$$

Therefore, the corresponding increase in entropy after the gas diffusion process is

$$\Delta S = S_1 - S_0 = -Nk(P_1 \ln P_1 + P_2 \ln P_3) \tag{4.42}$$

where $P_1 < 1$, and $P_2 < 1$. Since

$$I = \frac{\Delta S}{k \ln 2} = -N(P_1 \log_2 P_1 + P_2 \log_2 P_2) \tag{4.43}$$

it corresponds to information *loss* in this diffusion process.

It is interesting to compare the information measure provided by Shannon [1–3]:

$$I = -\sum_{j=1}^{N} P_j \log_2 P_j \tag{4.44}$$

For a *binary* information source, Shannon's formula is

$$I = -(P_1 \log_2 P_1 + P_2 \log_2 P_2) \tag{4.45}$$

Moreover, for *N independent binary* sources, the formula becomes

$$I = -N(P_1 \log_2 P_1 + P_2 \log_2 P_2) \tag{4.46}$$

Equation (4.46) is essentially the same as Eq. (4.43).

The interpretation of our example is that, after the diffusion process, it requires at least the amount of information I [Eq. (4.43)] to return the N_1 and N_2 atoms (if we could) into the original state. This amount of information is related to the amount of entropy increase [Eq. (4.42)]. In fact, this amount of information is related to the energy required to reverse the process [Eq. (4.25)], if we could.

For our third example, we use the second example—the diffusion problem—and include the effect due to the g factor in the entropy equation (4.26). Let us consider for a moment the quantity $Nk \ln g$, where $g = 2j + 1$. If we assume that the atoms of both gases have electronic spin $\frac{1}{2}$, for type 1 the electronic spin will be $\frac{1}{2}$ and for type 2 it will be $-\frac{1}{2}$. In the diffusion process, the atomic collisions make possible transition between the two kinds of atomic spins; the final state is assumed to be one with equiprobable spin orientations of the atoms. Thus this corresponds to $g = 2$, where $j = \frac{1}{2}$.

With the inclusion of the second example, after the diffusion process, the overall entropy equation becomes

$$S_2 = S_0 - Nk(P_1 \ln P_1 + P_2 \ln P_2 + \ln 2) \tag{4.47}$$

The overall increase in entropy is

$$\Delta S = S_2 - S_0 = -Nk(P_1 \ln P_1 + P_2 \ln P_2 + \ln 2) \tag{4.48}$$

This increase in entropy in the process is again recognized as information loss in the system:

$$I = \frac{\Delta S}{k \ln 2} = -N(P_1 \log_2 P_1 + P_2 \log_2 P_2 + 1) \tag{4.49}$$

In comparison with Eq. (4.43), there is a net change in information loss due to $g = 2$, which is equal to N bits, where N is the total number of atoms.

4.5 REMARKS

A low-entropy system is an unstable system. Eventually, it follows the normal evolutionary process leading to a more stable high-entropy state. From our examples, we see that information can be obtained from a physical system if and only if the entropy of the system progressively increases. In other words, a low-entropy system can be used to provide information, and is also known as a source of negentropy.

The information stored in a system can be regarded as a decrease in entropy of the system. The stored information is progressively destroyed or erased by the increase in entropy due to the normal evolution of the system. In practice it may take anywhere from a small fraction of a second to years, or even centuries, to erase the information, depending on the nature of the system (i.e., the recording material).

In statistical thermodynamics, entropy is defined as a measure of the disorder of a system. However, in information theory entropy is a measure of the *lack* of information about the actual structure of the system.

Historically, the idea of an entropy theory of information was first presented by Szilard [12, 13] in his famous paper, "On the Decrease of Entropy in a Thermodynamic System by the Intervention of Intelligent Beings," but at that time the concept was not well understood or accepted. It was not until the later 1940s that the relationship between entropy and information was rediscovered by Shannon [1–3].

Finally, for further reading on the entropy theory of information, we refer the reader to the excellent article and book by Brillouin [7, 8] and the papers by Rothstein [14, 15].

REFERENCES

1. C. E. Shannon, "A Mathematical Theory of Communication," *Bell Syst. Tech. J.*, vol. 27, 379–423, 623–656 (1948).
2. C. E. Shannon, "Communication in the Presence of Noise," *Proc. IRE*, vol. 37, 10 (1949).
3. C. E. Shannon and W. Weaver, *The Mathematical Theory of Communication*, University of Illinois Press, Urbana, 1962.
4. F. W. Sears, *Thermodynamics, the Kinetic Theory of Gases, and Statistical Mechanics*, Addison-Wesley, Reading, Mass., 1953.
5. M. W. Zemansky, *Heat and Thermodynamics*, 3rd ed, McGraw-Hill, New York, 1951.
6. P. G. Tait, *Sketch of Thermodynamics*, Edmonston and Douglas, Edinburgh, 1868, p. 100.
7. L. Brillouin, "The Negentropy Principle of Information," *J. Appl. Phys.*, vol. 24, 1152 (1953).
8. L. Brillouin, *Science and Information Theory*, 2nd ed, Academic, New York, 1962.
9. E. Schrödinger, *What is Life?*, Cambridge University Press, New York, 1945.
10. J. E. Mayer and M. G. Mayer, *Statistical Mechanics*, John Wiley, New York, 1940, p. 115.
11. A. Katz, *Principles of Statistical Mechanics*, W. F. Freeman, San Francisco, 1967, p. 97.

12. L. Szilard, "Über die Entropieverminderung in Einem Thermodynamishen System bei Eingriffen Intelligenter Wesen," *Z. Phys.*, vol. 53, 840 (1929).
13. L. Szilard, "On the Decrease of Entropy in a Thermodynamic System by the Intervention of Intelligent Beings" (translated by A. Rapoport and M. Knoller), *Behav. Sci.*, vol. 9, 301 (1964).
14. J. Rothstein, "Information, Measurement, and Quantum Mechanics," *Science*, vol. 114, 171 (1951).
15. J. Rothstein, "Information and Thermodynamics," *Phys. Rev.*, vol. 85, 135 (1952).

5
Demon Exorcist and Cost of Entropy

We discussed in detail the entropy theory of information in Chap. 4. The relationship between physical entropy and information theory was derived strictly from the statistical thermodynamic approach. We showed that information and entropy can be traded. However, information is gained inevitably at the expense of increasing entropy from a different source. This relationship of entropy and information was first recognized by Szilard [1, 2] in 1929 and later successfully applied by Brillouin [3, 4] in 1951. In practice, the amount of increasing entropy from a different source is generally excessive. In principle, the amount of information gain at best can only approach the equivalent amount of entropy traded off:

$$I \leq \frac{\Delta S}{k \ln 2} \tag{5.1}$$

In this chapter we discuss one of the most interesting applications of the entropy theory of information, namely, the demon's operation and information theory.

Of all the demons in the physical sciences, Maxwell's demon may be regarded as the most famous of all. The problem of Maxwell's demon and the theory of heat may be the best example for application of the entropy theory of information. What follows illustrates the impossibility of a perpetual motion machine. Subsequent sections demonstrate the demon's operation and information theory.

5.1 PERPETUAL MOTION MACHINE

According to the first law of thermodynamics, the law of energy conservation, a perpetual motion machine is impossible. Other perpetual motion machines, those of the second kind [5], used to convert the disorder of heat (e.g., by conduction or radiation) into work under the same temperature or pressure, are also known to be impossible, since they violate the second law of thermodynamics. For example, in a refrigerator the transfer of heat from a cold body to a hot body requires an external source of energy.

As illustrated in Chap. 4, a system can perform work if two bodies or subsystems with different temperatures or pressures are in contact. As an example, we denote these two bodies 1 and 2, and

$$T_2 > T_1 \tag{5.2}$$

where T_1 and T_2 are the temperatures of body 1 and body 2, respectively. When these two bodies are in contact, a certain amount of heat is expected to flow between them. Let the input heat into a body be a positive quantity; then we have

$$\Delta Q_2 = -K \Delta Q_1 \tag{5.3}$$

where K is a positive proportional constant, ΔQ_1 is the amount of heat input to body 1, a positive quantity, and ΔQ_2 is the amount of heat output from body 2, a negative quantity. Let W be the work done by the system, and then by the energy conservation law we have

$$\Delta Q_1 + \Delta Q_2 + \Delta W = 0 \tag{5.4}$$

Let us assume the thermodynamic process of the system is *reversible*; hence there is no increase in total entropy S:

$$\Delta S = \Delta S_1 + \Delta S_2 = 0 \tag{5.5}$$

By the definition of entropy Eq. (5.5) can be written

$$\frac{\Delta Q_1}{T_1} + \frac{\Delta Q_2}{T_2} = 0 \tag{5.6}$$

or, equivalently,

$$\Delta Q_2 = -\frac{T_2}{T_1} \Delta Q_1 \tag{5.7}$$

Thus

$$\Delta Q_1 + \Delta Q_2 = (T_1 - T_2) \Delta S_1 < 0 \tag{5.8}$$

which is a negative quantity. In Eq. (5.4) this amount equals the work done by the system. Therefore the system appears to perform work. The corresponding thermal efficiency [5] is

$$\eta = \frac{\Delta W}{\Delta Q_2} = \frac{T_2 - T_1}{T_2} < 1 \tag{5.9}$$

From Eq. (5.9), it can be seen that heat can be converted to work, but not completely. However, work can always be totally converted to heat. Thus from this heat-work transformation problem, it can be seen that it is not possible to realize a perpetual motion machine using a refrigerating system as an example. We illustrate in the following sections a perpetual motion machine of the second kind, from an information theory point of view. The information aspect of perpetual motion was first recognized by Szilard [1, 2] well before modern information theory was introduced by Shannon [6, 8] in 1948. However, it was not until Shannon's work appeared that the earlier work of Szilard was recognized and understood. We discuss these problems from the Maxwell's demon approach, but with the use of information theory. In short, a perpetual motion machine is again shown to be unrealizable in practice.

5.2 MAXWELL'S DEMON

The perpetual motion machine of the second kind was created by James Clerk Maxwell as early as 1871. The sorting demon, known as Maxwell's demon, first appeared in Maxwell's *Theory of Heat* [9, p. 328]. Since then it intrigued investigators of the physical world and has also provided an excellent example of the application of the entropy theory of information. Maxwell's sorting demon is some kind of intelligent being capable of tracing every molecule's motion. In other words he is able to perform certain tasks beyond the current physical constraints.

In order to illustrate the demon's operation, let us suppose we have a thermally insulated chamber filled with gas molecules. Let the chamber be divided into two parts, chamber 1 and chamber 2, as shown in Fig. 5.1. Also, let the partition wall be equipped with a small trapdoor which the demon is able to open and close as he wishes. The demon is able, we assume, to see every individual molecule, so he can allow a molecule to pass from one chamber to the other. Now the demon decides to let the fast molecules in chamber 1 pass into chamber 2, and the slower molecules in chamber 2 to pass into chamber 1. In this manner the demon is eventually able to collect all the fast molecules in chamber 2 and all the slower molecules in chamber 1. Thus, without any expenditure of work, he is able to raise the temperature in one chamber (chamber 2) and lower the temperature in the other. Hence work can be performed by the difference in temperature between the two chambers. Therefore we see that the demon is indeed able to create a perpetual motion machine of the second kind. This famous paradox has been explained by many physicists since the invention of Maxwell's demon. As a matter of fact, he has been known as a *temperature demon* in the theory of heat.

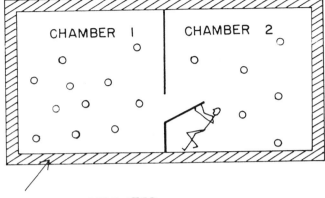

Fig. 5.1 Operation of Maxwell's demon.

There is a similar paradox involving *Brownian motion*, discussed by von Smoluchowski [10], in which a simple trapdoor device is used. The trapping procedure is based on the random opening and closing of a spring valve by the random bombardment of particles. In this manner, the spring valve creates high- and low-pressure chambers, so work can be performed, again, by pressure differences. This paradox was recognized by von Smoluchowski. Because of the random unpredictability and short duration of particle motion, he concluded that Brownian motion constitutes only an apparent contradiction to the second law of thermodynamics, so that permanent operation of such a system is not possible. That is, the system may move irregularly but not in a manner that allows partition of the particles. Thus a perpetual motion machine is again shown to be impossible.

A case similar to Brownian motion, but in which thermal agitation occurs in a rectifier, was also described by Brillouin [4]. He concluded that no rectified current can be produced by a physical rectifier. Once again the second law of thermodynamics is unchallenged.

Now we come to the application of information theory to the problem of Maxwell's demon. The basic question is: Are we sure the demon is actually able to see the individual molecules? This basic question was raised by Demers [11, 12] and Brillouin [3, 4], and the connection between information theory and the demon's operation was first fully discussed in their publications. We have adopted their basic explanation in the following discussion.

We again assume the problem of a thermally isolated chamber. Since the demon is within the enclosure, he is therefore in an equilibrium temperature state. The radiation within the chamber is only blackbody radiation; thus it is impossible for the demon to see anything inside the chamber. Raising the temperature of the chamber would not help, since the demon would be able to see only the thermal radiation and its random fluctuations, but not the molecules. Thus, under these isolated conditions, the demon is unable to operate the door. Separation of temperatures in the partitioned chamber is impossible.

However, one might ask: Although the demon cannot see the molecules, can he detect them by some other means? Yes he can. For example, he might detect the molecules by the electric field caused by the electric dipoles, or a magnetic field caused by the electronic spin, or even by their van der Waals forces. But all these fields are detectable only at a very short distance, since their field strength varies with the inverse square of the distance. Thus the demon can detect the presence of a molecule only when it is very close to the trapdoor. At this very short distance it is already too late for him to open the door without performing any work. Thus the demon must rely on detection of the molecule long before it approaches the trapdoor. Nevertheless, this long-distance detection requires a certain compensation for the external source of radiation. In our case, an optical radiator, (e.g., a flashlight) may be the simplest example. A detailed discussion of this example follows in Sec. 5.3.

5.3 INFORMATION AND DEMON EXORCIST

Let us now provide the demon with an electric flashlight to see the individual molecules within the isolated chamber. As pointed out in Chap. 4, the flashlight acts as an external source providing information for the demon. This is essentially what was called the source of negative entropy, or simply negentropy, by Brillouin [4, 13]. Then, by using the amount of information provided by the flashlight, the demon is able to operate the trapdoor. In this manner he is eventually able to decrease the entropy of the chamber to a lower state. Thus work can be performed by this lower entropy state. However, the entropy increase in the flashlight can be shown to be larger than or at least equal to the entropy decrease in the chamber. The total entropy (i.e., the combined entropy of the chamber and the flashlight) increases or remains constant, and as a result the second law of thermodynamics is still satisfied.

We now discuss in detail the problem of a temperature demon. We use the same thermally isolated chamber which is filled with gas molecules and in which the temperature is uniform throughout. Now we provide the demon with a flashlight with freshly charged batteries. In order to perceive the molecules the demon turns on the flashlight. The batteries heat up the filament of the light bulb to a temperature T_α higher than that of the gas:

$$T_\alpha > T_0 \tag{5.10}$$

where T_0 is the temperature of the gas. In order to obtain visible light within the chamber the flashlight is required to emit a quantum of light:

$$h\nu_\alpha > kT_0 \tag{5.11}$$

where h is Planck's constant, ν_α is the frequency of the visible light from the flashlight, k is Boltzmann's constant, and T_0 is the temperature of the gas in degrees Kelvin.

In a period of time, total energy ΔE is radiated from the flashlight. Then the corresponding amount of entropy increase is

$$\Delta S_\alpha = \frac{\Delta E}{T_\alpha} \tag{5.12}$$

This is also the reduction in entropy that the demon, *at best*, is able to obtain in the chamber by operating the trapdoor.

If the demon does not intervene, then the energy ΔE will be totally absorbed by the gas at temperature T_0. This corresponds to an increase in entropy in the gas

$$\Delta S_g = \frac{\Delta E}{T_0} \tag{5.13}$$

Now let us illustrate the demon's intervention. We have noticed that the demon is able to see an approaching molecule if and only if that molecule scatters at least a quantum of light $h\nu_\alpha$. The quantum of light is absorbed by the eye of the demon, or by means of a photodetector. This quantum of absorbed light represents an increase in entropy in the demon:

$$\Delta S_d = \frac{h\nu_\alpha}{T_0} \tag{5.14}$$

On the basis of the entropy theory of information Eq. (5.14) also provides the demon with an amount of information:

$$I_d = \frac{\Delta S_d}{k \ln 2} \tag{5.15}$$

Once the information is obtained, the demon can use it to decrease the entropy in the chamber. To compute the amount of entropy reduction, we let the initial entropy in the chamber be

$$S_0 = k \ln N_0 \tag{5.16}$$

where N_0 represents the total initial microscopic complexity of the gas. After receiving I_d, the demon is able to reduce the initial complexity N_0 to N_1:

$$N_1 = N_0 - \Delta N \tag{5.17}$$

where ΔN is the net change in complexity. Thus the final entropy in the chamber is

$$S_1 = k \ln N_1 \tag{5.18}$$

The corresponding amount of decreased entropy is

$$\Delta S_1 = S_1 - S_0 = k \ln\left(1 - \frac{\Delta N}{N_0}\right) \tag{5.19}$$

Since

$$\ln\left(1 - \frac{\Delta N}{N_0}\right) = -\left[\frac{\Delta N}{N_0} + \frac{1}{2}\left(\frac{\Delta N}{N_0}\right)^2 + \frac{1}{3}\left(\frac{\Delta N}{N_0}\right)^3 + \cdots\right] \tag{5.20}$$

Eq. (5.19) can be approximated by

$$\Delta S_1 \simeq -k\frac{\Delta N}{N_0} \qquad \text{for} \quad \frac{\Delta N}{N_0} \ll 1 \tag{5.21}$$

In conjunction with Eq. (5.14) one sees that the net entropy in the entire chamber is increased:

$$\Delta S = \Delta S_d + \Delta S_1 = k\left[\frac{h\nu_\alpha}{kT_0} - \frac{\Delta N}{N_0}\right] > 0 \tag{5.22}$$

since $h\nu_\alpha > kT_0$. It can be seen that the demon is not able to violate the second law of thermodynamics. All he can do is to convert only a small part of the total entropy increase in the flashlight into information; from this amount of information gain, he is able to reduce the entropy in the chamber. But the entropy reduction is much smaller than the cost in entropy incurred by the flashlight.

Let us proceed with the problem in more detail. We see that after a certain period the demon is able to establish different temperatures in the partitioned chamber:

$$\Delta T = T_2 - T_1 \tag{5.23}$$

where T_1 and T_2 are the corresponding temperatures in chambers 1 and 2. It can also be seen that

$$T_0 = T_2 - \frac{\Delta T}{2} = T_1 + \frac{\Delta T}{2} \tag{5.24}$$

In order for the demon to establish these two temperatures, he must divert the fast molecules from chamber 1 to chamber 2, and the slower molecules from chamber 2 to chamber 1. However, for the demon to be able to do this he needs at least two quanta of light: one for the fast molecule and another for the slower one. The two quanta of radiation must be provided by an external source, in our example, the flashlight. Therefore, the actual increase in entropy in the flashlight per two molecular exchanges is

$$\Delta S_d = 2 \frac{h\nu_\alpha}{T_0} \tag{5.25}$$

instead of Eq. (5.14). The corresponding amount of information gained by the demon is

$$I_d = \frac{2h\nu_\alpha}{T_0 k \ln 2} \tag{5.26}$$

If the kinetic energy of the faster and slower molecules is assumed to be

$$E_f = \tfrac{3}{2} k T_0 (1 + \varepsilon_1) \tag{5.27}$$

and

$$E_S = \tfrac{3}{2} k T_0 (1 - \varepsilon_2) \tag{5.28}$$

respectively, where ε_1 and ε_2 are arbitrary small positive constants, then the net transfer of energy due to these two molecules, from chamber 1 to chamber 2, and from chamber 2 to chamber 1, is

$$\Delta E = \tfrac{3}{2} k T_0 (\varepsilon_1 + \varepsilon_2) \tag{5.29}$$

Thus the corresponding entropy decrease per operation is

$$\Delta S_c = \Delta E \left(\frac{1}{T_2} - \frac{1}{T_1} \right) = -\tfrac{3}{2} k (\varepsilon_1 + \varepsilon_2) \frac{\Delta T}{T_0} \tag{5.30}$$

Thus the net entropy change in the entire process per operation is

$$\Delta S = \Delta S_d + \Delta S_c = k \left[\frac{2h\nu_\alpha}{kT_0} - \tfrac{3}{2}(\varepsilon_1 + \varepsilon_2) \frac{\Delta T}{T_0} \right] > 0 \tag{5.31}$$

where $(\varepsilon_1 + \varepsilon_2) \ll 1$, and $\Delta T \ll T_0$. Once again it is seen that violation of the second law of thermodynamics is impossible.

It would be interesting to calculate the efficiency of the demon. This can be obtained by using the ratio of the entropy decrease to the amount of information obtained by the demon:

$$\eta = -\frac{\Delta S_c}{I_d k \ln 2} = -\frac{\Delta S_c}{\Delta S_d} \qquad (5.32)$$

Therefore, by substituting Eqs. (5.25) and (5.30) into Eq. (5.32), we have

$$\eta = \frac{3k(\varepsilon_1 + \varepsilon_2)\Delta T}{4h\nu_\alpha} < 1 \qquad (5.33)$$

The demon's efficiency is *always* less than one; the most he can do is approach unit efficiency. However, for him to do so would require an extreme difference in temperature ($\Delta T \to \infty$).

One may question whether the problem of Maxwell's temperature demon can be rigorously explained, since occasionally we assume that $\Delta N \ll N_0$, $\Delta T \ll T_0$, and $\varepsilon_1 + \varepsilon_2 \ll 1$. However, we discuss Maxwell's pressure demon in Sec. 5.4 and illustrate a more rigorous demonstration of this problem.

We have achieved a significant result from an optical viewpoint in this section. Every observation must be compensated for by an increase in the entropy of another system. However this increase in entropy does not compensate for the decrease in entropy in the other system, so the second law of thermodynamics is not violated. Beyond this limit the observation (or use as a measurement) cannot be achieved in practice, since it basically violates the second law. The important result we have established is the basic connection of observation and information, which is presented in Chap. 6. This result also presents the idea that one cannot get something for nothing; there is always a price to pay. And there is always the question, "Can we afford it?" This is even truer in the information sciences—as Gabor [14] pointed out in his article, "Light and Information": "We cannot get something for nothing, not even an observation, far less a law of nature!"

Terminating this section, we emphasize that our treatment of the demon's effectiveness and information is primarily derived from the quantum condition, that is, $h\nu_\alpha$. A similar explanation can also be achieved by means of the classical approach. We refer the interested reader to the following articles: von Smoluchowski [10], Lewis [15], Slater [16], Demers [11,12], and, finally, the most interesting work by Brillouin [3,4,13], from whom we benefitted most.

5.4 DEMON EXORCIST, A REVISIT

In this section we discuss Maxwell's pressure demon. Instead of sorting out faster and slower molecules, this demon operates a trapdoor allowing the molecules to pass into one chamber but not in the opposite direction. Then, after a time, the demon is able to build up high pressure in one chamber and low pressure in the other. Thus work can be performed by the difference in pressure between the two connected chambers. One of the interesting facts under these conditions is that the demon does not even need to see the molecules.

In order to discuss the pressure demon's work, we adopt the model proposed by Brillouin [4], as shown in Fig. 5.2. We use a conduit to replace the trapdoor opening and to connect the two chambers. A beam of light from an external source is assumed to pass through a transparent section of the conduit near the trapdoor. The scattered light, if any, is used for detection of the molecules in the conduit, approaching the trapdoor. However, the appearance of molecules in the conduit is not sufficient for the demon to open the trapdoor, unless he knows the approximate average velocity of the molecules. In order for him to determine the molecular velocity, a set of revolving shutters is employed in the conduit, further away from the trapdoor, as shown in the figure. By using an appropriate shutter speed, it is possible to allow the passage of molecules having a certain average velocity. Molecules possessing either higher or lower velocities are reflected back into the chamber. Now with the photodetector and the revolving shutters, the demon is able to determine not only the presence of the molecules but also their approximate velocity.

We are now able to determine the amount of information the demon requires to operate the trapdoor. We let τ be a long interval of the pressure

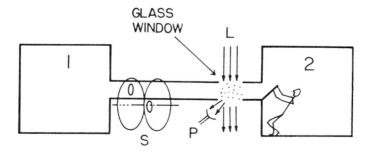

Fig. 5.2 Intervention by Maxwell's pressure demon. L, light beam; P, photodetector; S, rotating shutters.

demon's operation. He is now able to operate the trapdoor at every subinterval Δt_i of τ. To simplify, let every Δt_i be equal to Δt. At each Δt the demon needs to know whether any molecules are arriving from chamber 1 approaching the trapdoor at an average known speed. If there are, he will open the trapdoor to allow the molecules to pass through, and then immediately close it. If none are arriving, he will remain stationary. This problem is essentially a binary yes-or-no problem. Therefore the *a priori* probability of the binary source can be calculated for a long period of τ. Let Δt be sufficiently small so that the probability of opening the trapdoor more than once can be neglected. We denote by N_1 the total number of openings of the trapdoor, over τ, and by N_2 the total number of intervals during which the trapdoor remains closed, over τ. Then the corresponding *a priori* probabilities of the trapdoor opening or remaining closed at every Δt is

$$P_1 = \frac{N_1}{N} \tag{5.34}$$

and

$$P_2 = \frac{N_2}{N} \tag{5.35}$$

where $N = N_1 + N_2$, and $N = \tau/\Delta t$. Now the amount of information the demon needs is

$$I_d = -(P_1 \log_2 P_1 + P_2 \log_2 P_2) \qquad \text{bits}/\Delta t \tag{5.36}$$

To determine the *a priori* probabilities P_1 and P_2, we let $\bar{n}$ be the average number of molecules per Δt bombarding the trapdoor from chamber 1. Then $\bar{m} = \bar{n} Z_2 / Z_1$ is the average number of molecules from chamber 2 striking the trapdoor. Z_1 and Z_2 are the pressures in chamber 1 and chamber 2, respectively. From Appendix B, the *a priori* probabilities of P_1 and P_2 are

$$P_1 = 1 - e^{-\bar{n}} \tag{5.37}$$

and

$$P_2 = e^{-\bar{n}} \tag{5.38}$$

By substituting Eqs. (5.37) and (5.38) into Eq. (5.36), we have

$$I_d = -[(1 - e^{-\bar{n}}) \log_2(1 - e^{-\bar{n}}) + e^{-\bar{n}} \log_2 e^{-\bar{n}}] \tag{5.39}$$

Since Δt is assumed to be very small, it is also justifiable to assume that $\bar{n} \ll 1$. Therefore Eq. (5.39) can be approximated by

$$I_d \simeq \bar{n} \log_2\left(\frac{e}{\bar{n}}\right) \tag{5.40}$$

since $1 - e^{-\bar{n}} \simeq \bar{n}$, and $e^{-\bar{n}} \simeq 1 - \bar{n}$. The net (average) number of molecules entering chamber 2, per Δt, is

$$\Delta N = \bar{n} - \bar{m} P_1 \tag{5.41}$$

which is equal to the average number of molecules entering minus the average number of molecules leaving, per Δt. Since $\bar{n} \ll 1$, Eq. (5.41) can be approximated by

$$\Delta N \simeq \bar{n}\left(1 - \bar{n}\frac{Z_2}{Z_1}\right) \tag{5.42}$$

where Z_1 and Z_2 are the respective pressures of chambers 1 and 2. Thus the amount of entropy reduction per Δt accomplished by the demon is

$$\Delta S_c = -\Delta N k \ln\frac{Z_2}{N_1} = -k\bar{n}\left(1 - \bar{n}\frac{Z_2}{Z_1}\right)\ln\frac{Z_2}{Z_1} \tag{5.43}$$

But from Eq. (5.40) one can see that the lowest entropy increase in the external sources (i.e., the light source and the revolving shutter) is

$$\Delta S_d = I_d k \ln 2 \simeq k\bar{n} \ln\left(\frac{e}{\bar{n}}\right) \tag{5.44}$$

Therefore the net entropy change per Δt is

$$\Delta S = \Delta S_d + \Delta S_c = k\bar{n}\left[\ln\left(\frac{e}{\bar{n}}\right) - \left(1 - \bar{n}\frac{Z_2}{Z_1}\right)\ln\frac{Z_2}{Z_1}\right] > 0 \tag{5.45}$$

From Eq. (5.45) we see that the second law of thermodynamics has not been violated.

From Eqs. (5.43) and (5.44) the pressure demon's efficiency is

$$\eta = -\frac{\Delta S_c}{\Delta S_d} = \frac{[1 - \bar{n}(Z_2/Z_1)]\ln(Z_2/Z_1)}{\ln(e/\bar{n})} < 1 \tag{5.46}$$

which is always smaller than one. The best the demon can do is to make the efficiency approach unity but, in order for him to do so, there must be an extremely large pressure difference between the chambers ($Z_2/Z_1 \to \infty$).

In concluding this section, we emphasize that, in order for Maxwell's demons (i.e., the temperature and pressure demons) to operate a perpetual motion machine, a certain amount of information must be obtained and must be derived from some external source for which the entropy is increased. As a result of the information gained by the demon, the entropy of the thermodynamic process is reduced and work can be performed. However, the entropy decrease created by the demon is still *smaller*, even under ideal conditions, than the amount of information provided by the external sources. Therefore, even with this most powerful Maxwell's demon, from an information theory point of view, the second law of thermodynamics has been proved to remain unchallenged.

5.5 SZILARD'S DEMON

In the earlier days, before Shannon's information theory, most physicists thought that information had nothing to do with the physical laws. However, the basic relationship between physical entropy and information was discovered by Szilard [1, 2] as early as 1929 and is recorded in one of his remarkable works. Unfortunately, this interesting work was neglected for many years, and not until the development of information theory by Shannon [6–8] in 1948 was it brought to light.

The work on the entropy theory of information by Szilard is, to our knowledge, one of the earliest on this subject. It is certainly the most interesting and rigorous demonstration involving Maxwell's demon and information theory. Because of its importance and historical significance in relation to the entropy theory of information, we devote a separate section of this chapter to Szilard's machine.

Szilard's machine consists of a cylinder of volume V and a frictionless piston acting as a partition wall dividing the cylinder into two chambers, 1 and 2, as shown in Fig. 5.3. As the piston moves either to the left or the right, it raises one of the pans and lowers the other. For convenience, the piston is equipped with a trapdoor which can be opened or closed. Szilard assumed that the cylinder contains only *one* molecule, and that the whole cylinder maintains a constant temperature T. The molecule continuously gains and loses kinetic energy as it randomly bounces off the walls, and it has an average kinetic energy of $3kT/2$. When the trapdoor is opened, no work is performed by simply sliding the piston sideways.

Now let us assume that the piston is moved to a position near the center of the cylinder. In this location, the demon clamps the trapdoor shut, and at this moment he somehow knows where the molecule is located. Let us suppose that the molecule is in chamber 1. Then the frictionless piston moves slowly from left to right, as a result of the random movement of the molecule against the piston walls. In this manner, work is performed by Szilard's machine. After chamber 1 has expanded to the whole volume of the cylinder, the demon opens the trapdoor and slides the piston back

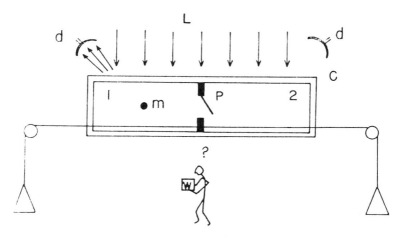

Fig. 5.3 Szilard's machine with intervention of the demon. d, photodetector; L, light beams; C, cylinder; b, pan; w, weight; P, piston; m, molecule.

to its original position. In this fashion another cycle of work is done by the machine. Thus a perpetual motion machine of the second kind can be realized through the demon's intervention. But a puzzle similar to the previous one remains—how can the demon locate the molecule?

As in the previous examples, the demon needs to obtain a certain amount of information in order to locate the molecule. From this information he is able to reduce the entropy of the system, but the amount of information provided will be shown to be in excess of the entropy reduction.

To illustrate the demon's work, we let V_1 and V_2 be the respective volumes of chambers 1 and 2. The first step is to locate the molecule. For simplicity, we let the cylinder be transparent. We equip the demon with two beams of light, one through chamber 1 and the other through chamber 2, as shown in Fig. 5.3. If light is scattered from one of these chambers, it will show where the molecule is located. However, the absorption of the light quanta hv by one of the photodetectors represents an entropy increase in the process:

$$\Delta S_c = \frac{hv}{T} > k \tag{5.47}$$

where $hv > kT$, h is Planck's constant, v is the frequency of the light beams, T is the Kelvin temperature in the cylinder, and k is Boltzmann's constant. It was noted by Szilard [1, 2] that a slightly lower entropy limit can be calculated:

$$\Delta S_c \geq k \ln 2 = 0.7k \tag{5.48}$$

We can now compute the amount of information the demon needs for its intervention. Let P_1 and P_2 be the *a priori* probability of locating the molecule in chamber 1 and in chamber 2, respectively:

$$P_1 = \frac{V_1}{V} \tag{5.49}$$

and

$$P_2 = \frac{V_2}{V} \tag{5.50}$$

where $P_1 + P_2 = 1$ and $V_1 + V_2 = V$.

Now the amount of information required by the demon is

$$I_d = -(P_1 \log_2 P_1 + P_2 \log_2 P_2) \tag{5.51}$$

which comes from a binary information source. For every cycle of the demon's operation, this much information is needed in order for the system's entropy to be reduced. The largest amount of information needed must correspond to the equiprobable case:

$$P_1 = P_2 = \tfrac{1}{2} \tag{5.52}$$

which also implies that

$$V_1 = V_2 = \tfrac{1}{2} V \tag{5.53}$$

Then the largest amount of information required by the demon is

$$I_{max} = 1 \text{ bit/cycle} \tag{5.54}$$

The demon can use this bit of information to reduce the system's entropy:

$$\Delta S_d = -k \ln 2 \tag{5.55}$$

By comparing Eqs (5.47) and (5.55), we see that the net entropy change per cycle of operation is

$$\Delta S = \Delta S_c + \Delta S_d \geq 0 \tag{5.56}$$

Again the result is within the constraints of the second law of thermodynamics.

Now let us apply the entropy equation for an ideal monatomic gas

[Eq. (4.26)]. This equation can be simplified for our present purpose to the case of one molecule, $N = 1$:

$$S = K + k \ln V \tag{5.57}$$

where

$$K = \frac{5k}{2} + k \ln \left[g \left(\frac{4\pi m E}{3h^2} \right)^{3/2} \right]$$

is a positive constant. At the moment the molecule is located, the entropy decrease can be computed. For example, if the molecule is in chamber 1,

$$\Delta S_1 = k \ln \left(\frac{V_1}{V} \right) = k \ln P_1 \tag{5.58}$$

and, if the molecule is in chamber 2,

$$\Delta S_2 = k \ln \left(\frac{V_2}{V} \right) = k \ln P_2 \tag{5.59}$$

Hence the average entropy decrease per cycle is

$$\Delta S_d = P_1 \Delta S_1 + P_2 \Delta S_2 = -I_d k \ln 2 \tag{5.60}$$

where $I_d \leq 1$; the equality holds for $V_1 = V_2$. Therefore, for $V_1 \neq V_2$, the net entropy change per cycle of the demon's intervention is

$$\Delta S = \Delta S_c + \Delta S_d = k \left(\frac{h\nu}{kT} - I_d \ln 2 \right) > 0 \tag{5.61}$$

In concluding this section we have demonstrated examples of Maxwell's demon in the application of modern information theory. We have shown that the receiving of information by the demon is inevitably accompanied by a certain compensatory (with some excess) increase in the entropy of some external source or system. This increase in the entropy traded for information is one of the basic relationships between information theory and the physical science.

5.6 DIFFRACTION-LIMITED DEMON

The sorting demon of Maxwell has intrigued scientists for some time. Maxwell's demon is an excellent example in the entropy theory of information. Since we are in the computer epoch, we will extend Maxwell's demon to a diffraction-limited demon, for which we assume that the demon has a diffraction-limited eye, as shown in Fig. 5.4. Since the demon is located in a thermally isolated chamber, in order for him to see the molecules, we

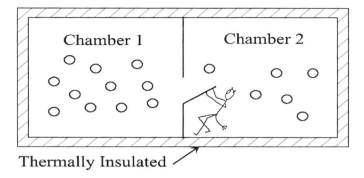

Fig. 5.4 Diffraction-limited demon exorcist.

have provided him with a laser. By using the negentropy provided by the laser, the demon is able to let the molecules go through the trapdoor. In this manner, he is eventually able to decrease the entropy of the chamber from a higher entropy state to a lower state. However, a question is raised: Is the demon actually able to see the individual molecule since his eye is diffraction limited? If he is not able to resolve the individual molecule, then how can he be so sure that he is letting only the right molecule pass through the trapdoor? In order for him to do so, we equip the demon with a powerful computer. To perceive the arriving molecules, the demon turns on the laser, which is required to emit at least a quanta of light for him to make the observation, that is,

$$h\upsilon = kT \tag{5.62}$$

We assume that the quanta of light reflected by the approaching molecules is totally absorbed by the eye of the demon, which corresponds to an increase of entropy in the demon, that is

$$\Delta S_d = \frac{h\upsilon}{T} \tag{5.63}$$

which is equivalent to the amount of information provided to the demon, that is,

$$I_d = \frac{\Delta S_d}{k \ln 2} \tag{5.64}$$

Because of the diffraction-limited nature of the demon's eye, the demon needs to process the absorbed quanta to a higher resolution, so that he is able to resolve the molecules and to allow the passages of the high-

or low-energy molecules through the trapdoor. The amount of information gain, through the processing by the equipped computer, constitutes an equivalent amount of entropy increased, as given by

$$\Delta S_p = k \Delta I_d \ln 2 \tag{5.65}$$

where ΔI_d is the incremental amount of information provided by the computer. With this amount of information gain, the demon is able to reduce the entropy of the chamber to a lower state.

To compute the amount of entropy reduction by the demon's intervention, we let the initial entropy of the chamber be

$$S_0 = k \ln N_0 \tag{5.66}$$

where N_0 is the initial microscopic complexity of the chamber. After receiving I_d from the laser, and ΔI_d from the computer, the demon is able to reduce the initial complexity from N_0 to N_1,

$$N_1 = N_0 - \Delta N \tag{5.67}$$

where ΔN is the net changed in complexity. The final entropy of the chamber is, therefore

$$S_1 = k \ln N_1 \tag{5.68}$$

for which the net entropy reduction of the chamber is

$$\Delta S_1 = S_1 - S_0 = k \ln\left(1 - \frac{\Delta N}{N_0}\right) \approx -k \frac{\Delta N}{N_0} \tag{5.69}$$

where $\Delta N \ll N_0$.

However, we see that the overall net entropy changed in the chamber, per trapdoor operation by the demon, would be

$$\Delta S = \Delta S_d + \Delta S_p + \Delta S_1 = k\left(\frac{h\upsilon}{kT} + \Delta I_d \ln 2 - \frac{\Delta N}{N_0}\right) > 0 \tag{5.70}$$

in which we see that the diffraction limit demon's exorcist is still within the limit of the second law of thermodynamics, where $h\upsilon > kT$, $\Delta I_d > 0$, and $\Delta N \ll N_0$.

5.7 MINIMUM COST OF ENTROPY

One question still remains unanswered: What would be the minimum cost of entropy required for the demon to operate the trapdoor? Let the arrival molecules at the trapdoor at an instant be one, two, or more molecules. Then the computer is required to provide the demon with a "yes" or a "no" information. For example, if a single molecule is approaching the trapdoor (say, a high-velocity one), the demon will open the trapdoor to allow the molecule to go through the trapdoor. Otherwise, he will stand still.

For simplicity, if we assume a 50 percent probability of one molecule arriving at the trapdoor, then the demon needs one additional bit of information, from the computer, for him to open the trapdoor. This additional bit of information, however, corresponds to the amount of entropy increase provided by the computer, i.e.,

$$\Delta S_p = k \ln 2 \approx 0.7k \tag{5.71}$$

which is the minimum cost of entropy required for the demon to operate the trapdoor. This minimum cost of entropy is in fact the result obtained by Szilard's demon in Sec. 5.5. From Eq. (5.70) we see that the overall net entropy increased in the chamber is

$$\Delta S = k\left(\frac{h\upsilon}{kT} + 0.7 - \frac{\Delta N}{N_0}\right) > 0 \tag{5.72}$$

If one takes into account the other bit of "no" information, provided by the computer, then the average net entropy increased in the chamber per operation would be

$$\Delta S_{\text{ave}} = k\left(\frac{2h\upsilon}{kT} + 1.4 - \frac{\Delta N}{N_0}\right) \gg 0 \tag{5.73}$$

in which two quantas of light radiation are required. It is trivial, if one includes the possibility of a slower molecule approaching the trapdoor, the average cost of entropy per operation even higher.

Even though we omit the two quantas of light in the calculation, the overall net entropy change in the chamber is still increasing, i.e.,

$$\Delta S_{\text{ave}} = k\left(1.4 - \frac{\Delta N}{N_0}\right) > 0 \tag{5.74}$$

where $\Delta N \ll N_0$. In other words, the entropy compensated by the computer is still higher than the entropy reduced by the demon. In view of this argument, we see that the computer we provided for the demon is also operated within the limit of the second law of thermodynamics.

We should now discuss the cost of entropy required to increase the resolution beyond diffraction limit. Let a classical imaging system having the following resolution limit:

$$r_m = \frac{1.22\lambda f}{D} \tag{5.75}$$

where r_m is the minimum separation, λ is the wavelength of the light source, and f and D are the focal length and the diameter of the imaging aperture. To

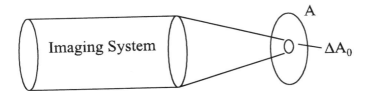

Fig. 5.5 Observation error: A is the field of view and ΔA_0 is the observation error.

increase the resolution of the system, one could either reduce the wavelength or enlarge the aperture of the system, or both. However, if the wavelength reduction and the aperture enlargement are not the subject to be considered, then the observed image can be processed by the computer beyond the resolution limit. However, the amount of resolution gain would be compensated by the amount of entropy increased. Since the minimum cost of entropy to resolve a point-object is $0.7k$, for n object points resolution the minimum cost of entropy would be $0.7nk$. Nevertheless, in practice, the actual cost of entropy is very excessive.

For example, let A be the field of view of an imaging system and ΔA_0 the observation error, as shown in Fig. 5.5; then the amount of information obtained by this observation would be

$$I = \log_2 \frac{A}{\Delta A_0} \tag{5.76}$$

where $\Delta A_0 = \pi(r_m)^2$ and r_m is the minimum resolvable distance of the optical system. If the observed image is then processed to a higher resolution, then a net amount of information gain is anticipated. In other words, the observation error of the imaging system can be reduced, by processing the observed image from ΔA_0 and ΔA_1. The net information gain provided by the computer is

$$\Delta I = \log_2 \frac{\Delta A_0}{\Delta A_1} \qquad \Delta A_1 \ll \Delta A_0 \tag{5.77}$$

Thus, we see that in the minimum cost of entropy is

$$\Delta S = \Delta I k \ln 2 = k \ln 2 \log_2 \frac{\Delta A_0}{\Delta A_1} \tag{5.78}$$

where ΔA_0 and ΔA_1 are the initial and final observation errors.

In short, we have achieved a very fundamental conclusion: One cannot get something from nothing; there is always a price to pay. Resolution beyond the resolution limit is possible, but only through the increase of entropy from a certain physical system (in our example, the computer).

However, in practice, the cost of entropy is very excessive. In fact, the additional information gain at best can only approach to the equivalent amount of entropy traded off:

$$\Delta I \leq \frac{\Delta S}{k \ln 2} \tag{5.79}$$

5.8 GABOR'S PERPETUUM MOBILE OF THE SECOND KIND

We now discuss Gabor's [14] perpetual motion machine of the second kind. Gabor's problem essentially concerns a modified version of Szilard's machine, but without an intelligent demon's intervention. However, it relies on a certain relay activation; thus the major difference in Gabor's problem is that the operation is accomplished strictly by physical devices.

In order to show that Szilard's machine operates within the limits of the second law of thermodynamics, we had to assume that such an observation by the demon cannot be obtained without a certain increase in entropy of some other source. In fact, this is one of the basic relationships between observation and information, which we elaborate on in Chap. 6.

Now we come to an illustration of Gabor's perpetuum mobile, as pictured in Fig. 5.6. Let us assume that a single thermally agitated molecule is wandering within a cylinder of volume V at a constant temperature T. It is also assumed that the lower part of the cylinder is transparent and that a beam of light illuminates this lower chamber, with a volume of V_1, as shown in the figure. The light beam is intended to provide information on the location of the wandering molecule, and is derived from a heated filament. For simplicity, we assume a monochromatic source of frequency v. As shown in Fig. 5.6, a set of ideal reflectors (mirrors) is used to keep the light beam circulating through the lower chamber without any loss.

Now if the molecule wanders within the lower chamber of the cylinder, certain scattered light from it is expected. As in the previous examples, this scattered light can be detected by means of photodetectors. A relay is then activated in such a way that a frictionless piston is immediately inserted into the cylinder, and two sliding mirrors then cover the transparent portion. As a result of the random bombardment of the piston wall by the molecule, work can be performed by the slow expansion of the lower chamber. The corresponding entropy decrease after the molecule has been trapped in the lower chamber is therefore

$$\Delta S_d = k \ln \frac{V_1}{V} \tag{5.80}$$

where $V_1 < V$.

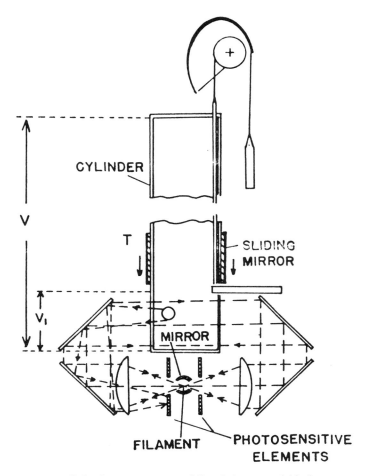

Fig. 5.6 Gabor's perpetuum mobile of the second kind.

When the piston has moved to the extreme end of the cylinder ($V_1 = V$) it again slides out of the cylinder and returns to its original location. In the same manner, another cycle of operation can take place. If the work done by the machine is greater than the overall energy lost from the light source, then a perpetual motion machine of the second kind has been achieved—this time without help from a demon. Here we see again that, in order to activate the relay, scattered light from the molecule is required. So we again come to the basic question of how much scattered light is needed for such an operation. Or, equivalently, how small is the amount of information required for such motion?

In order to answer this basic question, let t_0 be the average total amount of time consumed during every cycle of detection; Δt is the duration of the scattered light, and $\Delta t < t_0$. It can be seen that t_0 and Δt are directly proportional to the volumes of the cylinder V and V_1, respectively. Let us recall the information measure

$$I = \log_2 \frac{N_0}{N_1} \tag{5.81}$$

where N_0 and N_1 are the respective initial and final number of equiprobable stages. We also noted in our problem that

$$\frac{N_0}{N_1} = \left(\frac{t_0}{\Delta t}\right)\left(\frac{V}{V_1}\right) = \left(\frac{V}{V_1}\right)^2 \tag{5.82}$$

Therefore the smallest amount of information required to activate the relay is

$$I = 2\log_2 \frac{V}{V_1} \tag{5.83}$$

From Eq. (5.65) it can be seen that

$$Ik\ln 2 + \Delta S_d = k\ln\frac{V}{V_1} > 0 \tag{5.84}$$

which is about twice the entropy decrease.

Since the information provided for the detection is derived from the increase in entropy from an external source—in our case the light source—the expenditure of the entropy is excessive. It is a simple matter to show that the overall entropy increase is far greater than the amount of entropy the system can reduce:

$$\Delta S - Ik\ln 2 > 0 \tag{5.85}$$

Thus

$$\Delta S + \Delta S_d > 0 \tag{5.86}$$

From the above discussion, it is clear that Gabor's perpetuum mobile still operates within the limits of the second law of thermodynamics.

REFERENCES

1. L. Szilard, "Über die Entropieverminderung in Einem Thermodynamischen System bei Eingriffen Intelligenter Wesen," *Z. Phys.*, vol. 53, 840 (1929).
2. L. Szilard, "On the Decrease of Entropy in a Thermodynamic System by the Intervention of Intelligent Beings" (translated by A. Rapoport and M. Knoller), *Behav. Sci.*, vol. 9, 301 (1964).

3. L. Brillouin, "Maxwell's Demon Cannot Operate. Information Theory and Entropy I," *J. Appl. Phys.*, vol. 22, 334 (1951).
4. L. Brillouin, *Science and Information Theory*, 2nd ed., Academic, New York, 1962.
5. F. W. Sear, *Thermodynamics, the Kinetic Theory of Gases, and Statistical Mechanics*, Addison-Wesley, Reading, Mass., 1953.
6. C. E. Shannon, "A Mathematical Theory of Communication," *Bell Syst. Tech. J.*, vol. 27, 379–423, 623–656 (1948).
7. C. E. Shannon, "Communication in the Presence of Noise," *Proc. IRE*, vol. 37, 10 (1949).
8. C. E. Shannon and W. Weaver, *The Mathematical Theory of Communication*, University of Illinois Press, Urbana, 1949.
9. J. H. Jeans, "Dynamical Theory of Gases," 3rd ed., Cambridge University Press, New York, 1921, p. 183.
10. M. von Smoluchowski, "Experimentell Nachweisbare, der Üblichen Thermodynamik Widersprechende Molekular-Phänomene," *Phys. Z.*, vol. 13, 1069, (1912).
11. P. M. Demers, "Les Démons de Maxwell et le Second Principle de la Thermodynamique," *Can. J. Res.*, vol. 22, 27 (1944).
12. P. M. Demers, "Le Second Principle et la Théorie des Quanta," *Can. J. Res.*, vol. 23, 47 (1945).
13. L. Brillouin, "The Negentropy Principle of Information," *J. Appl. Phys.*, vol. 24, 1152 (1953).
14. D. Gabor, "Light and Information," in E. Wolf, Ed., *Progress in Optics*, vol. I, North-Holland, Amsterdam, 1961.
15. G. N. Lewis, "The Symmetry of Time in Physics," *Science*, vol. 71, 569 (1930).
16. J. C. Slater, *Introduction to Chemical Physics*, McGraw-Hill, New York, 1939.
17. F. T. S. Yu, "Superresolution and cost of entropy," *Optik*, vol. **54**, 1–7 (1979).

6
Observation and Information

In the classical theory of light, an observation can be made as small as we please. However, for some very small objects this assumption does not hold true. For example, let us take a Mach-Zehnder interferometer [1], as shown in Fig. 6.1. It can be seen that only a very small fraction of light is required to illuminate the object transparency $s(x, y)$. This small fraction of light, however, carries all the information that we intended. However, the other path of light carries almost all the energy of the light source. The corresponding irradiance distributed on the photographic emulsion is [2]

$$I(p, q) = R^2 + |S(p, q)|^2 + 2RS(p, q)\cos[\alpha_0 p + \phi(p, q)] \tag{6.1}$$

where R is the background (reference) beam, $S(p, q)$ is the corresponding Fourier spectrum of the object transparency, (p, q) are the corresponding spatial frequency coordinates, α_0 is an arbitrary constant, and

$$S(p, q) = |S(p, q)| \exp[i\phi(p, q)] \tag{6.2}$$

In our example, it is assumed that $R \gg |S(p, q)|$. Thus the interference term (the last term) is considered very small in comparison with the background irradiance. In principle, it is possible to enhance the weak interference term by means of a coherent optical processor [2], so it is simply a matter of making $s(x, y)$ observable. In the classical theory, it is possible to observe with as weak an object beam as we wish. However, we know instinctively that this is not practical, since an increase in background light would at the same time increase background fluctuation. Thus the informational term (the interference term), could be completely buried in the background fluctuation. So in practice there exists a practical lower limit. Beyond this limit it is not possible to retrieve the information (the object).

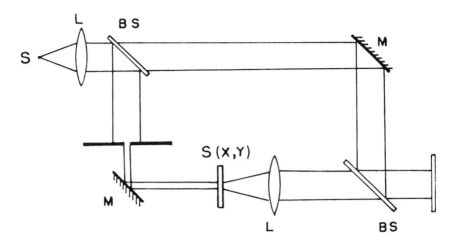

Fig. 6.1 The paradox of observation without illumination. S, Monochromatic point source; M, mirror; $s(x,y)$, object transparency; P, photographic plate; BS, beam splitter; L, lens.

In our case, the informational term must be at least equal to or greater than the mean-square fluctuation of the background irradiance:

$$R^2 \langle S^2(p,q) \rangle \geq \langle (\delta R^2)^2 \rangle \tag{6.3}$$

where $\langle \ \rangle$ denotes the spatial ensemble, which implies that

$$\langle S^2(p,q) \rangle \geq \varepsilon_0 \tag{6.4}$$

with ε_0 the lower limit of irradiance.

Substituting the equality in the above equation in Eq. (6.3) yields

$$\frac{R^2}{\varepsilon_0} = \frac{\langle (\delta R^2)^2 \rangle}{\varepsilon_0^2} \tag{6.5}$$

This is the minimum background irradiance, in terms of ε_0, that allows the observation to be made. As Gabor [3, 4] noted, Eq. (6.5) is Poisson's *law of rare events*. This was what was accounted for in the hypothesis that monochromatic light arrives in quanta. In our case these quanta of light arrive in a random fashion and are subjected to constraints; that is, on the average, the background irradiance R^2/ε_0 arrives during the observation (the recording).

Thus no observation is possible if less than *one quantum* of light arrives from the object transparency. This simple statement essentially leads us to the concept of the quantum theory of light, namely, the monochromatic light field is perceived in *discrete quanta*.

In Chap. 5 we reached a significant conclusion: An observation can take place only when there is a certain compensatory increase in the entropy of some external source. This result is regarded as the basic relationship between observation and information theory, which we discuss here. We utilize several important results which were demonstrated by Brillouin [5–7].

6.1 OBSERVATION WITH RADIATION

Since light is electromagnetic in nature, it is the purpose of this section to discuss the probabilistic analysis of observations made by radiation. We note that, as the radiation frequency increases, the quantum effect takes place. When the quantum $h\nu$ is no longer a small quantity as compared with kT, a photodetector takes the quantum effect into account:

$$E_n = nh\nu \qquad n = 1, 2, 3, \ldots$$

where the E_n's are the quantized energy levels.

These energy levels at a temperature in Kelvins have the following probability distribution (see Appendix C), known as *Gibb's distribution*:

$$P_n = Ke^{-nh\nu/kT} \tag{6.6}$$

with

$$\sum_{n=0}^{\infty} P_n = 1 \tag{6.7}$$

where

$$K = 1 - e^{-h\nu/kT}$$

Thus Eq. (6.6) can be conveniently written

$$P_n = e^{-nh\nu/kT} - e^{-(n+1)h\nu/kT} \tag{6.8}$$

The mean quantum state can be determined by

$$\bar{n} = \sum_{n=0}^{\infty} nP_n = \frac{1}{e^{h\nu/kT} - 1} \tag{6.9}$$

It is noted that, if $h\nu \ll kT$, then

$$\bar{n} \simeq \frac{kT}{h\nu} \qquad \text{for } h\nu \ll kT \tag{6.10}$$

which is the average energy of the photodetector:

$$\bar{E} = \bar{n}h\nu = kT \tag{6.11}$$

In the observation analysis, we denote the probabilities

$$P(0 \leq n < g) = \sum_{n=0}^{g-1} P_n = 1 - e^{-gh\nu/kT} \tag{6.12}$$

and

$$P(n \geq g) = \sum_{n=g}^{\infty} P_n = 1 - P(0 \leq n < g) = e^{-gh\nu/kT} \tag{6.13}$$

where g is an arbitrary quantum state.

Now let us observe the photodetector. Let the quantum state g be used as a threshold decision level for the observation. On the photodetector, at time t, energy corresponding to $n \geq g$ quanta is called a *positive observation* (reading). It can be seen that the probability of making an observation error is $P(n \geq g)$, primarily because of the thermal fluctuation of the photodetector. However, the probability of making a correct observation is $P(0 \leq n < g)$, since it is basically due to the absorption of additional quanta from some external source which are carried over the g threshold level.

We can now compute an optimum decision level for the observation. In order to do so, we must first compute the corresponding *median quantum state*. By definition, let us equate either Eq. (6.12) or (6.13) to $\frac{1}{2}$, which yields

$$m = \frac{kT}{h\nu} \ln 2 \tag{6.14}$$

where m denotes the median quantum state. Since $h\nu \ll kT$, we see from Eq. (6.10) that

$$m = \bar{n} \ln 2 < \bar{n} \tag{6.15}$$

which is smaller than the average quantum state $\bar{n}$. The corresponding *median energy* of the photodetector is

$$E_m = mh\nu = kT \ln 2 \tag{6.16}$$

From Eq. (6.11), we see that

$$E_m < \bar{E} \tag{6.17}$$

As we see later, the situation is entirely reversed for high-frequency observation.

We now shall calculate the average quantum stages for the cases of $0 \leq n < g$, and $n \geq g$. Let us define the following reduced sample space probabilities:

$$P(n/0 \leq n < g) = \frac{P_n}{P(0 \leq n < g)} \tag{6.18}$$

and

$$P(n/n \geq g) = \frac{P_n}{P(n \geq g)} \tag{6.19}$$

Then the corresponding average quantum states are

$$\bar{n}_1 = \sum_{n=0}^{g-1} nP(n/0 \leq n < g) = \bar{n} - \frac{g}{e^{ghv/kT} - 1} \tag{6.20}$$

and

$$\bar{n}_2 = \sum_{n=g}^{\infty} nP(n/n \geq g) = \bar{n} + g \tag{6.21}$$

It is also noted that

$$\bar{n} = \bar{n}_1 P(0 \leq n < g) + \bar{n}_2 P(n \geq g) \tag{6.22}$$

For an optimum decision observation, we adopt the median m for the threshold level, $g = m$, and then Eqs. (6.20) and (6.21) yield, respectively,

$$\bar{n}_1 = \bar{n} - \frac{m}{e^{mhv/kT} - 1} \tag{6.23}$$

and

$$\bar{n}_2 = \bar{n} + m \tag{6.24}$$

It is noted that $hv \ll kT$. Then Eqs. (6.23) and (6.24) can be written

$$\bar{n}_1 \simeq \bar{n}(1 - \ln 2) = 0.3\bar{n} \tag{6.25}$$

and

$$\bar{n}_2 = \bar{n}(1 + \ln 2) = 1.7\bar{n} \tag{6.26}$$

If one accepts that $g = m$ is the threshold decision level for the observation, then we have an error probability of 50% per observation. However, if we choose a higher quantum state for the decision level ($g > m$), then a more reliable observation can be obtained. For example, if $g = 4m$ is chosen, then the error probability of the observation will be about 2%.

It is interesting to note from Eq. (6.21) that the average energy above the g quantum state is

$$\bar{E}_2 = \bar{n}_2 h\nu = \bar{n} h\nu + gh\nu \tag{6.27}$$

where $gh\nu$ can be regarded as the *excess* energy. This excess energy can come from two possible sources, either from an internal thermal fluctuation or from an external source. If it emanates purely from an external source, then g quanta were actually absorbed in the photodetector and added to the normal average energy of $\bar{n} h\nu$. But if this excess energy is purely from an internal thermal fluctuation, then it will be compensated for by a loss in energy at a later instant in time. Thus the photodetector has an average energy of kT. However, if it is from an external source, then the excess energy of $gh\nu$ will eventually be dissipated into the surrounding environment of T in the photodetector. This corresponds to an entropy increase in the photodetector:

$$\Delta S = \frac{gh\nu}{T} \tag{6.28}$$

We see that, if $g = m$ is chosen for the decision level, then Eq. (6.28) becomes

$$\Delta S = k \ln 2 \tag{6.29}$$

since $m = kT \ln 2 / h\nu$. Equation (6.29) can be regarded as the *minimum cost* (the optimum) in entropy needed for an observation, with a 50% chance of error. In terms of the amount of information, Eq. (6.29) is essentially equal to 1 bit:

$$I = \frac{\Delta S}{k \ln 2} = 1 \text{ bit} \tag{6.30}$$

This is the amount of information obtained per observation by the photodetector. In fact, Eq. (6.29) is a consequence of Szilard's [8, 9] results mentioned in Chap. 5. We emphasize that the result of Eq. (6.29) is not restricted only to the low-frequency case; it can also be applied to high-frequency observation.

We can now discuss high-frequency observation, and whether the quantity of $h\nu$ is higher or lower in the order of kT. In this case $h\nu/kT$ is no longer a small quantity. Let us first state the quantity:

$$h\nu \gg kT \tag{6.31}$$

We know that the smallest amount of energy the photodetector can absorb is one quantum. In view of $h\nu \gg kT$ and the 50% chance of error criterion ($g = m$), $m = 1$. From Eq. (6.15) we see that

$$m = \bar{n} = 1 \tag{6.32}$$

and with Eq. (6.14) we now have

$$hv = kT \ln 2 \tag{6.33}$$

Since the total probability for the higher quantum states $(n > m)$ is 50%, then the probability of the ground state is also 50%. Hence if we observe one quantum or more in the photodetector, the chance that the absorption is from an external source is 50%, and the chance that it is caused by thermal fluctuation is also 50%. Now if one quantum of absorption (the minimum) is due to an external source, then this corresponds to an entropy increase in the photodetector:

$$\Delta S = k \ln 2 \tag{6.34}$$

which is identical to Eq. (6.29) for low-frequency observation and corresponds to a minimum cost in entropy in high-frequency observation.

In view of $hv \gg kT$, the minimum cost we pay for high-frequency observation is even greater:

$$\Delta S > k \ln 2 \tag{6.35}$$

However, since $m < 1$, we use $g = 1$, an integer, which provides us with a more reliable observation. Of course, this is better than a 50% chance of error, and the amount of information obtained is still the same, exactly 1 bit. This is the minimum information required for the functioning of Szilard's demon but the cost is much higher when this method is used.

We emphasize that this information is available only for a very short instant in time. The excess energy (or rather the entropy) provided by the external source is eventually quietly dissipated in the photodetector. This excess energy dissipation provides the observation within the limits of the second law of thermodynamics.

Thus far our discussion has been restricted to only the ensemble point of view. That is, for a special observation, the minimum cost may be lower than $k \ln 2$, but under no circumstances have we been able to predict when this will occur. Thus only in terms of statistical averages will we be provided with a logical answer. However, our discussion has revealed the basic relationship between information theory and observation; we use this basic relationship to further explore observation and information.

6.2 SIMULTANEOUS OBSERVATIONS

In this section we consider an interesting example of many simultaneous observations. We see that the minimum cost in entropy for simultaneous observations is even higher than $k \ln 2$.

Let us assume that we are simultaneously observing α photodetectors. The observation can be performed with one, or with more than one, of these α detectors, so that the energy is seen to be greater than an arbitrary threshold level of the detector. For simplicity, let us assume that all these photodetectors are identical and are maintained at the same Kelvin temperature T. First we consider the low-frequency observation, $hv \ll kT$, and follow this immediately with the high-frequency observation, $hv \gg kT$,. Let us recall the probabilities of Eqs. (6.12) and (6.13) and denote them, respectively, P_1 and P_2:

$$P_1 = P(0 \leq n < g) = 1 - e^{-E_0/kT} \tag{6.36}$$

and

$$P_2 = P(n \geq g) = e^{-E_0/kT} \tag{6.37}$$

where $E_0 = ghv$, a threshold energy level.

Since there are α photodetectors, the probability that β detectors deflect energy below E_0, at an instant in time, is

$$P_{\alpha,\beta} = \frac{\alpha!}{\beta!(\alpha - \beta)!} P_1^\beta P_2^{\alpha-\beta} \tag{6.38}$$

which is a *binomial* probability distribution.

We first assume that an observation gives a positive result if any one of the α photodetectors gives rise to an energy level equal to or above E_0. Therefore the probability of a correct observation is P_1^α, and $1 - P_1^\alpha$ corresponds to the observation error. If we impose a 50% chance of error constraint on the α simultaneously observed detectors, we will then have

$$P_1^\alpha = (1 - e^{-E_0/kT})^\alpha = \tfrac{1}{2} \tag{6.39}$$

It can be seen that, if $\alpha = 1$, Eq. (6.39) is reduced to the previous result:

$$E_0 = kT \ln 2 \qquad \text{for } \alpha = 1 \tag{6.40}$$

From Eq. (6.39), one can determine that E_0 is a function of α:

$$E_0 = -kT \ln[1 - (\tfrac{1}{2})^{1/\alpha}] > 0 \tag{6.41}$$

where α is the number of photodetectors. It can be seen that E_0 increases as the number of detectors α increases. The lower limit of E_0 corresponds to $\alpha = 1$, as in Eq. (6.40).

If we take the natural logarithm of Eq. (6.39), we will have

$$\alpha \ln(1 - e^{-E_0/kT}) = -\ln 2 \tag{6.42}$$

We see that E_0 increases as α increases. By the logarithmic expansion

$$\ln(1-x) = -\left(x + \frac{x^2}{2} + \frac{x^3}{3} + \cdots\right) \quad \text{for } x^2 < 1 \tag{6.43}$$

Eq. (6.42) can be approximated:

$$\alpha e^{-E_0/kT} \simeq \ln 2 \tag{6.44}$$

which corresponds to large values of α and E_0. Thus Eq. (6.41) can be further reduced to

$$E_0 \simeq kT[\ln \alpha - \ln(\ln 2)] \quad \text{for } \alpha \gg 1 \tag{6.45}$$

Since the additional energy from the external source eventually dissipates in the photodetectors, it corresponds to an increase in entropy:

$$\Delta S = k(\ln \alpha + 0.367) > k \ln 2 \tag{6.46}$$

where $\ln(\ln 2) = -0.367$.

Equation (6.46) indicates the minimum cost in entropy for *one* of the positive readings from the α photodetectors. It is also clear that additional energy from the external source is absorbed by *all* the other photodetectors. Since this energy is assumed to be less than the threshold energy level E_0, the actual cost in entropy is even higher:

$$\Delta S > k(\ln \alpha + 0.367) \tag{6.47}$$

By a similar analysis, one can show that, for γ simultaneous positive readings (correct observations) from α photodetectors, the corresponding net entropy increase is

$$\Delta S = \gamma k(\ln \alpha + 0.367) \tag{6.48}$$

where $1 < \gamma < \alpha$. From the same account of low-energy absorption for the remaining $(\alpha - \gamma)$ detectors, we conclude that

$$\Delta S > \gamma k(\ln \alpha + 0.367) \tag{6.49}$$

The net entropy increase in all the detectors is even higher than proportional to γ.

With this basic analysis for low-frequency observations, we can now carry the analysis to the high-frequency case, $h\nu \gg kT$.

As noted earlier a single quantum of $h\nu$ is sufficient for a reliable observation (reading), since in practice high-frequency oscillation does not exist in the photodetector (a blackbody radiator). Hence we simply take the threshold energy level

$$E_0 = h\nu > kT \ln 2 \quad \text{for } h\nu \gg kT$$

for the case of only *one* reading from the α photodetectors. By the same reasoning, as in the low-frequency observations, the net entropy increase is

$$\Delta S > \frac{h\nu}{T} > k \ln 2 \tag{6.50}$$

Furthermore, for γ positive observations out of the α simultaneous observations, the net entropy increase is

$$\Delta S > \frac{\gamma h\nu}{T} > \gamma k \ln 2 \tag{6.51}$$

In concluding this section, we point out that, for many simultaneous observations, the minimum cost in entropy is higher than that for a single observation. The cost in entropy is generally proportional to the number of simultaneous positive observations. As for high-frequency observation, the cost in entropy is even higher, since $h\nu \gg kT$. But a more reliable observation in the high-frequency case can generally be achieved. Furthermore, it is also emphasized that the case of $h\nu \simeq kT$ should be analyzed separately and under different special conditions.

6.3 OBSERVATION AND INFORMATION

We now come to the basic problem of observation and information. As we have stressed, the amount of information obtained from an observation is ultimately derived from some compensatory increase in the entropy of other sources. We have also demonstrated, with physical observations, that no one has been able to claim a perfect observation. Thus certain observation errors are bound to occur.

In order to emphasize the notion of information as applied to an optical observation, let us define the information measure for the observation [10]:

$$I = \log_2 \frac{N_0}{N_1} \quad \text{bits} \tag{6.52}$$

where N_0 is the number of equiprobable states *before* the observation, and N_1 is the number of equiprobable states *after* the observation.

Let us designate the total field of view A, an optical system. For simplicity, the optical system is composed of compound lenses. We observe a position at a certain point in A, say in a (x, y) spatial domain, with an observation error of ΔA, as shown in Fig. 6.2 ΔA is the spatial resolution limit of the optical system. The amount of information in this observation

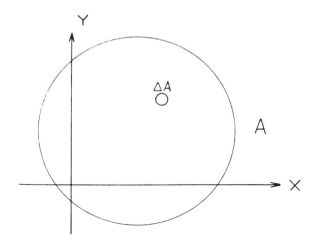

Fig. 6.2 Experimental error observation: A is the total field of view, and ΔA is the observation error at position (x,y).

can be obtained from the ratio of the initial uncertainty of A to the final uncertainty of ΔA:

$$I = \log_2 \frac{A}{\Delta A} \quad \text{bits} \tag{6.53}$$

Accordingly, the accuracy of the observation can be defined as [6]

$$\mathcal{A} \triangleq \frac{A}{\Delta A} \tag{6.54}$$

Thus the amount of information obtained by the observation can also be written

$$I = \log_2 \mathcal{A} \quad \text{bits} \tag{6.55}$$

In application, let the resolving power of the optical system be [2, 7, 10]

$$\Delta A_i = \pi \left(\frac{1.22 \lambda f}{D}\right)^2 \tag{6.56}$$

where ΔA_i is the remaining uncertainty of the observation. The subscript i refers to the image plane, λ is the wavelength of the light source, f is the focal distance, and D is the diameter of the optical aperture.

By substituting Eq. (6.56) into Eq. (6.53), the amount of information obtained is

$$I = \log_2 \frac{A_i}{\pi} \left(\frac{D}{1.22 \lambda f}\right)^2 \quad \text{bits} \tag{6.57}$$

where A_i is the total field of view, which refers to the image plane. From the above equation, we see that the amount of information increases as the size of the optical aperture increases.

It is interesting to note that, if the optical system makes several completely different observations, then the total amount of information obtained is

$$I_0 = \sum_{n=1}^{N} I_n \quad \text{bits} \tag{6.58}$$

where

$$I_n = \log_2 \frac{A_{in}}{\pi} \left(\frac{D}{1.22 \lambda f}\right)^2 \tag{6.59}$$

and N denotes the total number of different observations. However, if the N observations are completely repetitive (redundant), then the total information obtained is no more than that of a single observation:

$$I_0 = I_n \tag{6.60}$$

Thus, with N sequential observations, one expects that

$$I_n \leq I_0 \leq \sum_{n=1}^{N} I_n \tag{6.61}$$

We now discuss several simultaneous observations made by several optical systems. We see that, if these optical systems see different fields of view, then the amount of information provided is equal to the sum. However, if the fields are not completely different, then the amount of information provided is somewhat smaller than the sum. If the fields of view are identical, then the amount of information provided is equal to that of any one of the channels.

Now let us discuss the optical spatial channels. Let N be the number of independent spatial channels with channel capacities of $C_1, C_2, \ldots, C_N$, as shown in Fig. 6.3. The overall spatial capacity is then

$$C = \sum_{n=1}^{N} C_n \tag{6.62}$$

Observation and Information

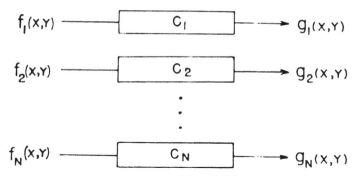

Fig. 6.3 Block diagram of the input-output spatial channels.

where the C_n are defined as the upper limit of the spatial mutual information [11]:

$$C_n \triangleq \max I_n(f, g) \qquad (6.63)$$

where $f(x, y)$ and $g(x, y)$ are the input-output spatial signals ensemble, and n denotes the nth spatial channel. Thus the overall spatial capacity is the sum of the maximum mutual information.

6.4 ACCURACY AND RELIABILITY IN OBSERVATIONS

We discuss problems of observation within a space-and-time domain. This involves observing a position on a spatial plane, at an instant in time, with a certain observation error. It is emphasized that the spatial domain of observation must be bounded; otherwise the information cannot be defined, since the information provided would lead to an infinite amount.

In order to illustrate the accuracy of observation, we assume a spatial domain A, which corresponds to the total field of view of an optical system. The spatial domain A is then subdivided into small areas ΔA, which are limited by the resolvable power of the optical system. Of course, light is necessary for the observation, in which a particle or object is assumed to be wandering within the spatial domain A. In practice, we look at each subdivision ΔA until we locate the particle. The accuracy of observation, as defined in Eq. (6.54), is

$$\mathcal{A} \triangleq \frac{A}{\Delta A} = \alpha \qquad (6.64)$$

where α is the total number of ΔA's within A. To look for the particle we simply illuminate each ΔA by a beam of light, and each ΔA is assumed to be equipped with a photodetector able to detect scattered light, if any, from the particle. As in previous examples, we assume that each of the photodetectors is maintained at a constant Kelvin temperature T. Later we discuss low-accuracy observation which corresponds to $h\nu \ll kT$. We also discuss separately sequential and simultaneous observations.

Let us start with sequential observations. We investigate each of the succeeding photodetectors until a positive reading is obtained, say from the qth ΔA, where $q \leq \alpha$. The reading may be caused by thermal fluctuation in the detector but, if it happens to be a positive reading, then the particle has been found in one of the ΔA's out of the q possibilities. Hence the amount of information obtained by this sequential observation is

$$I = \log_2 q \quad \text{bits} \tag{6.65}$$

Since the positive reading was obtained from the absorption of scattered light by the qth photodetector then, according to Eq. (6.41), the accompanying entropy increase in the qth detector is

$$\Delta S \geq -k \ln[1 - (\tfrac{1}{2})^{1/q}] \qquad q \geq 1 \tag{6.66}$$

For a large value of q, the right-hand side of Eq. (6.66) can be approximated [see Eq. (6.47)]:

$$\Delta S \geq k(\ln q + 0.367) > k \ln 2 \tag{6.67}$$

Thus

$$\Delta S - Ik \ln 2 \geq 0.367k > 0 \tag{6.68}$$

which is a positive quantity.

For the case of simultaneous observations, we assume that we have observed γ positive readings simultaneously from the α photodetectors. The amount of information obtained is therefore

$$I = \log_2 \frac{\alpha}{\gamma} \quad \text{bits} \tag{6.69}$$

where $\gamma \leq \alpha$. Again, we see that any reading could be due to thermal fluctuation. Since there are γ detectors absorbing the scattered light, and the observations are made on all the α photodetectors, the overall amount of entropy increase in the γ photodetectors is

$$\Delta S \geq -\gamma k \ln[1 - (\tfrac{1}{2})^{1/\alpha}] \tag{6.70}$$

For a large value of α, Eq. (6.70) can be approximated as

$$\Delta S \geq \gamma k (\ln \alpha + 0.367) \tag{6.71}$$

which increases with respect to γ and α. Thus

$$\Delta S - Ik \ln 2 \geq k [\ln \gamma + (\gamma - 1) \ln \alpha + 0.367\gamma] > 0 \qquad (6.72)$$

It is interesting to note that, if it takes only one of the α photodetectors to provide a positive reading ($\gamma = 1$), then Eq. (6.72) is essentially identical to Eq. (6.68). However, if $\gamma \gg 1$, then the amount of information obtained from the simultaneous observations [Eq. (6.69)] is somewhat less than that obtained from the sequential observations of Eq. (6.65), for $q \gg 1$, and the amount of entropy increase for the simultaneous observations is also greater.

Since it is assumed that only one particle is wandering in the spatial domain A, for $\gamma = 1$, Eq. (6.72) yields the smallest tradeoff of entropy and information. At the same time, for a large number of photodetectors ($\alpha \gg 1$), Eq. (6.72) is the asymptotic approximation used in high-accuracy observation ($h\nu \gg kT$). It is also emphasized that any other arrangement of the photodetectors may result in higher entropy. For example, if all the photodetectors are arranged to receive light directly, rather than from scattered light, then it can be seen that a high entropy cost will result.

We now discuss high-accuracy observation. It is noted that, if ΔA becomes very small, then higher-frequency illumination (a shorter wavelength) is necessary for the observation. As illustrated in Fig. 6.4, this

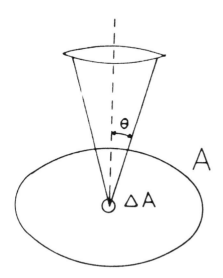

Fig. 6.4 A high-accuracy observation.

observation cannot be efficient unless the wavelength of the light source is shorter than $1.64d \sin \theta$ (see Ref. 2):

$$\gamma \leq 1.64d \sin \theta \tag{6.73}$$

where d is the diameter of ΔA, and θ is the subtended half-angle of the lens aperture. Accordingly,

$$d = \frac{1.22\lambda}{2 \sin \theta} \tag{6.74}$$

is the well-known formula for resolving power, where $2 \sin \theta$ is the *numerical aperture*. Thus the frequency required for the observation must satisfy the inequality

$$v = \frac{c}{\lambda} \geq \frac{c}{1.64d \sin \theta} \tag{6.75}$$

where c is the speed of light.

Now let us use the lower limit of Eq. (6.75) for a definition of the characteristic diameter or distance of the detector, which is assumed to maintain a constant temperature T:

$$\frac{hv}{kT} = \frac{hc}{1.64kTd \sin d} = \frac{d_0}{1.64d \sin \theta} \tag{6.76}$$

The characteristic diameter d_0 is therefore

$$d_0 = \frac{hc}{kT} \simeq \frac{1.44}{T} \tag{6.77}$$

Thus for high-frequency observation, such as $hv \gg kT$, we see that

$$d \ll d_0 \tag{6.78}$$

However, for low-frequency observation, such as $hv \ll kT$, we see that

$$d \gg d_0 \tag{6.79}$$

It is emphasized that d_0 possesses no physical significance except that, at a given temperature T, it indicates the boundary between low- and high-frequency observations.

Let us recall Eq. (6.41) for high-frequency observation:

$$hv > E_0 = -kT \ln[1 - (\tfrac{1}{2})^{1/\alpha}] \tag{6.80}$$

where E_0 is the threshold energy level for the photodetectors. For $\alpha \gg 1$, Eq. (6.80) can be approximated by

$$hv > kT(\ln \alpha + 0.367) \tag{6.81}$$

Since the absorption of one quantum is adequate for a positive reading, the corresponding entropy increase is

$$\Delta S = \frac{hv}{T} > k(\ln \alpha + 0.367) \tag{6.82}$$

From Eq. (6.69) we obtain the information

$$I = \log_2 \alpha \quad \text{bits} \tag{6.83}$$

Therefore we conclude that

$$\Delta S - Ik \ln 2 > 0.367k > 0 \tag{6.84}$$

Except for the equality, this ΔS is identical to that of the low-frequency observation of Eq. (6.68). However, the entropy increase is much higher, since v is very high. We emphasize that, the higher the frequency used, the finer the observation obtained. This is the price we pay in entropy for greater accuracy of observation.

One must discuss the case of $hv \simeq kT$ separately and with special attention. For instance, if Eq. (6.70) holds, it does not necessarily mean that its asymptotic approximation [Eq. (6.71)] also holds.

We now come to the reliable observation. One must distinguish the basic difference between accuracy and reliability in observations. From the viewpoint of statistical communication [12–14], a reliable observation is directly related to the chosen decision threshold level E_0. That is, the higher the threshold level, the greater the reliability. However, accuracy in observation is inversely related to the spread of the pulse signal: The narrower the spread, the greater the accuracy. A simple example illustrating the difference between these two concepts is shown in Fig. 6.5. It is evident that, the higher the threshold energy level E_0 chosen, the greater the reliability of observation. However, higher reliability also corresponds to higher probability of a miss. If the decision level E_0 is set at a lower level, then a less reliable observation is expected. Thus higher probability of error (a false alarm) results, because of thermal fluctuation (noise). It is noted that, in *decision theory* [13, 14], given an *a priori* probability of noise, the cost of a miss, and the cost of a false alarm, an optimum decision level can be found.

Now instead of considering only a 50% chance of error in observation, we equate the error probability to the inverse of the reliability [6]:

$$P(n \geq g) = \frac{1}{\mathcal{R}} \tag{6.85}$$

where $\mathcal{R}$ is the reliability. We consider first the reliability in low-frequency

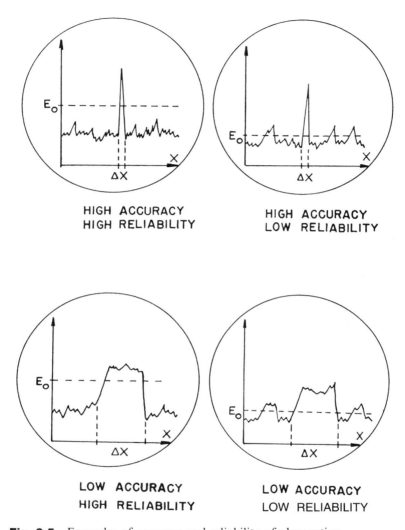

Fig. 6.5 Examples of accuracy and reliability of observation.

observation, $h\nu \ll kT$. It is also noted that low-frequency observation corresponds to low-accuracy observation. Let us use the error probability of Eq. (6.13) to determine the reliability:

$$P(n \geq g) = e^{-gh\nu/kT} = \frac{1}{\mathcal{R}} \tag{6.86}$$

Observation and Information

which can be written

$$\ln \mathcal{R} = \frac{gh\nu}{kT} \tag{6.87}$$

From Eq. (6.27), we see that after the observation the excess energy $gh\nu$ is eventually dissipated in the photodetector; thus it corresponds to an increase in entropy:

$$\Delta S \geq \frac{gh\nu}{T} = k \ln \mathcal{R} \tag{6.88}$$

It is also interesting to consider the reliability in the problem of many simultaneous observations. Let us recall Eq. (6.39), the probability of a correct observation when simultaneously observing α photodetectors. We have

$$(1 - e^{-gh\nu/kT})^\alpha = 1 - \frac{1}{\mathcal{R}} \tag{6.89}$$

It can be seen that, for $\alpha = 1$ (the case of a single photodectector), Eq. (6.89) is identical to Eq. (6.87). Alternatively, Eq. (6.89) can be written

$$\alpha \ln(1 - e^{-gh\nu/kT}) = \ln\left(1 - \frac{1}{\mathcal{R}}\right) \tag{6.90}$$

From Eq. (6.43), we determine that, for large values of α and $\mathcal{R}$, the left- and right-hand sides of Eq. (6.90) can be approximated by

$$\alpha \ln(1 - e^{-gh\nu/kT}) \simeq -\alpha e^{-gh\nu/kT} \tag{6.91}$$

and

$$\ln\left(1 - \frac{1}{\mathcal{R}}\right) \simeq -\frac{1}{\mathcal{R}} \tag{6.92}$$

Thus we have

$$\alpha e^{-gh\nu/kT} \simeq \frac{1}{\mathcal{R}} \tag{6.93}$$

or

$$\ln(\alpha \mathcal{R}) \simeq \frac{gh\nu}{kT} \quad \text{for } \alpha \mathcal{R} \gg 1 \tag{6.94}$$

Since Eq. (6.94) is identical to Eq. (6.87), for $\alpha = 1$, we see that Eq. (6.94) is a good approximation, even for a low value of α. Since it is assumed that there are γ correct observations from the α detectors, the entropy increase is

$$\Delta S \geq \gamma \frac{gh\nu}{T} = \gamma k \ln(\alpha \mathcal{R}) \tag{6.95}$$

We now discuss high-frequency observation, $hv \gg kT$. As noted previously, one quantum hv per photodetector is adequate for an observation, since $hv \gg kT$. Thus the entropy increase is [see Eq. (6.51)]

$$\Delta S \geq \gamma \frac{hv}{T} \tag{6.96}$$

Again, for intermediate-frequency observation, $hv \simeq kT$, the reliability analysis must be treated separately in each case.

We now consider the efficiency of observation. As defined in Chap. 5, the efficiency of observation is

$$\eta = \frac{Ik \ln 2}{\Delta S} \tag{6.97}$$

In the low-frequency case ($hv \ll kT$), the efficiency of many simultaneous observations, for $\gamma = 1$ [from Eqs. (6.69) and (6.71)], is

$$\eta = \frac{\ln \alpha}{\ln \alpha + 0.367} = \frac{1}{1 + 0.367/\ln \alpha} \simeq 1 - \frac{0.367}{\ln \alpha} \tag{6.98}$$

It can be seen that, the higher α, and the higher the accuracy $\mathcal{A}$, the higher the efficiency of observation. Thus the efficiency approaches unity as $\alpha \to \infty$.

It would be interesting to determine the efficiency of observation as a function of reliability. Again, for simplicity, we let $\gamma = 1$; therefore the efficiency is [from Eqs. (6.83) and (6.95)]

$$\eta = \frac{Ik \ln 2}{\Delta S} = \frac{\ln \alpha}{\ln (\alpha \mathcal{R})} = \frac{1}{1 + \ln \mathcal{R}/\ln \alpha} \tag{6.99}$$

Thus we see that the observation efficiency depends on the ratio of logarithmic reliability to logarithmic simultaneous observations, α. The higher this ratio is, the lower the observation efficiency.

In the case of high frequency, $hv \gg kT$, the efficiency of observation can be obtained from Eqs. (6.83) and (6.96). Thus we have

$$\eta = kT \frac{\ln \alpha}{hv} \tag{6.100}$$

Since $hv > kT(\ln \alpha + 0.367)$, the efficiency is lower than that in the low-frequency case.

In concluding this section, we point out that high-frequency observation corresponds to high accuracy and high reliability. However, the observation efficiency is somewhat lower than that in the low-frequency case.

6.5 OBSERVATION BY INTERFERENCE AND BY MICROSCOPE

In Sec. 6.4 we discussed accuracy and reliability in observation. We illustrated with examples the amount of entropy traded for the amount of information obtained and found it relatively small. We saw that from Eq. (6.98)—at least for relatively low reliability—one is able to push the observation efficiency close to unity. From Eq. (6.99) we concluded that observation efficiency can be high for relatively high reliability and high accuracy. However, efficiency is generally somewhat less in high-frequency observation than in the low-frequency case. With a relatively high cost in entropy in high-frequency observation, both high accuracy and high reliability can be achieved.

In this section we discuss and illustrate two other methods of observation, namely, interference and the microscope. We use these two methods to show that the amounts of entropy involved are indeed great, and that the observation efficiencies are even lower than in the previous examples.

It is common knowledge in the field of interferometry that the separation between two reflecting walls or two particles can be obtained by means of an interferometric technique. For example, if one wishes to determine the distance between two particles, one observes or measures the interference fringes between the two particles, as shown in Fig. 6.6. It is clear that these interference fringes, resulting from standing waves between them, can be obtained simply by illuminating the particles with a monochromatic light source, as shown in the figure. Here we see that the half-wavelength of the light source is smaller than the separation of the two particles:

$$\frac{\lambda}{2} \leq x_1 \qquad (6.101)$$

where x_1 is the distance between the two particles. By counting the number of interference fringes, one can determine the separation x_1 between the particles:

$$x_1 = \frac{\alpha \lambda}{2} \qquad (6.102)$$

where α is the number of interference fringes.

If it is assumed that the observation is accomplished by scanning a tiny photodetecting probe between the particles (providing the perturbation of the field is negligible), then the detector requires g quanta hv in order for it to respond properly at the location of maximum interference. Since

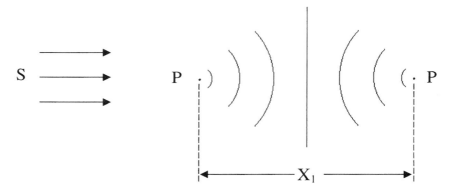

Fig. 6.6 Observation by interference. S, monochromatic plane wave; P, particle; X_1, separation.

there are α fringes in each scanning, the total amount of energy required is

$$\Delta E \geq \alpha g h \nu \tag{6.103}$$

The corresponding amount of entropy increase in the photodetector per scanning is

$$\Delta S \geq \frac{\Delta E}{T} = \frac{\alpha g h \nu}{T} \tag{6.104}$$

Let us define the characteristic distance [see Eq. (6.77)]:

$$\Delta x_0 = \frac{hc}{kT} \simeq \frac{1.44}{T}$$

where c is the speed of light. Then, according to Eq. (6.102), the frequency of the light source can be written

$$\nu = \frac{c}{\lambda} = \frac{\alpha c}{2x_1} \tag{6.105}$$

Thus the entropy increase can be written

$$\Delta S \geq \frac{\alpha k g \, \Delta x_0}{2 \, \Delta x} \tag{6.106}$$

where $\Delta x = x_1/\alpha$, the separation between fringes. From Eq. (6.106) we can determine the minimum number of quanta g required for the interference observation. We first discuss low-frequency interference observation. This corresponds to a larger separation between the two particles, as compared with the wavelength λ.

Now for the low-frequency case, $hv \ll kT$, we have

$$\frac{hv}{kT} = \frac{\alpha hc}{2kTx_1} = \frac{\Delta x_0}{2\Delta x} \ll 1 \qquad (6.107)$$

In interference observation, it is quite analogous to the problem of many simultaneous observations, discussed in Sec. 6.2. Thus for a 50% error probability (reliability $\mathcal{R} = 2$), we have the decision threshold level [Eq. (6.41)]:

$$E_0 = ghv = -kT \ln[1 - (\tfrac{1}{2})^{1/\alpha}] \qquad (6.108)$$

Therefore the minimum number of quanta required for the observation per scanning is

$$g = -\frac{kT \ln[1 - (\tfrac{1}{2})^{1/\alpha}]}{hv} \qquad (6.109)$$

By substituting Eq. (6.109) into Eq. (6.104), the entropy increase is found:

$$\Delta S \geq -\alpha k \ln[1 - (\tfrac{1}{2})^{1/\alpha}] \qquad (6.110)$$

It is also clear that the amount of information per scanning, using the interference observation method is

$$I = \log_2 \frac{x_1}{\Delta x} = \log_2 \alpha \qquad (6.111)$$

which is essentially the same as Eq. (6.83). Thus the efficiency of the interference observation can be determined:

$$\eta = \frac{Ik \ln 2}{\Delta S} = \frac{\ln \alpha}{-\alpha \ln[1 - (\tfrac{1}{2})^{1/\alpha}]} \qquad (6.112)$$

If one compares Eq. (6.112) with Eq. (6.98), it is evident that the efficiency is much smaller here than in the previous low-frequency cases. Particularly when α is large, Eq. (6.112) is approximated by

$$\eta \simeq \frac{\ln \alpha}{\alpha(\ln \alpha + 0.367)} < 1 \qquad (6.113)$$

Again we see that

$$\Delta S - Ik \ln 2 \geq k\{-\alpha \ln[1 - (\tfrac{1}{2})^{1/\alpha}] - \ln \alpha\} > 0 \qquad (6.114)$$

Now let us discuss the high-frequency case, $hv \gg kT$. As noted, high-frequency observation corresponds to high-accuracy observation. In interference observation it also corresponds to shorter-distance observation. Thus we have

$$\frac{hv}{kT} = \frac{\Delta x_0}{\Delta x} \gg 1 \qquad (6.115)$$

Here the high-frequency interference observation is well beyond the limit of blackbody radiation of the photodetector which is maintained at Kelvin temperature T. As noted earlier, in high-frequency observation one quantum per fringe detection is sufficient; thus the total amount of energy required per scanning observation is

$$\Delta E \geq \alpha h v = \frac{\alpha h c}{2\,\Delta x} \tag{6.116}$$

The corresponding entropy increase in the photodetector is

$$\Delta S \geq \frac{\alpha h v}{T} = \frac{\alpha h c}{T 2\,\Delta x} \tag{6.117}$$

Moreover, from Eq. (6.117), we see that

$$\Delta E\,\Delta x \geq \frac{\alpha h c}{2} \tag{6.118}$$

which bears a resemblance to the uncertainty relation. However, the physical significance is quite different from that of the uncertainty relation. That is, in the uncertainty relation (discussed in Sec. 6.5) ΔE corresponds to the error in energy, while in the interference observation ΔE is the minimum energy required for the observation. It is noted that ΔE of the interference observation is at least partially dissipated into heat during the observation.

Furthermore, from Eq. (6.118) we see that, as Δx becomes infinitesimally small, the minimum required energy for observation increases to an infinite amount. However, in practice, there exists a physical limit. As Δx approaches a certain limit, the Heisenberg uncertainty principle applies. The observation cannot be obtained without introducing errors. This subject of uncertainty and observation is discussed in greater detail in Sec. 6.6.

Now let us discuss observation under a microscope. In this problem a large number of photons is required. The amount of entropy required in microscopic observation is great in comparison to the previous cases.

For simplicity, let us assume that a circular wave guide of radius r_0 is used to focus the range of observation, as shown in Fig. 6.7. The light propagated through the waveguide has discrete wavelengths (quanta) with which to satisfy the boundary conditions. Superimposing these discrete quanta of the light field results in a focus spot, say at $Z = Z_0$. Let us then suppose a circularly polarized transverse electric wave exists in the wave guide, for which the wave equation is

$$\nabla^2 H_z = \frac{1}{c^2}\frac{\partial^2 H_z}{\partial t^2} \tag{6.119}$$

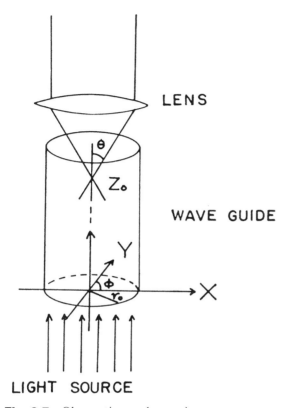

Fig. 6.7 Observation under a microscope.

where ∇^2 is the Laplacian operator, and c is the speed of light. The corresponding magnetic vector is [15, 16]:

$$H_z = \sum_{n=1}^{\infty} H_{0n}(\cos n\phi + i \sin n\phi) J_n(ur) \cos(\omega t - \gamma z) \qquad (6.120)$$

where the H_{0n} are arbitrary constants, the $J_n(ur)$ are Bessel functions of the first kind, γ is a propagation constant, $\omega = 2\pi\nu$, and ν is the frequency of the light source. The corresponding cutoff wavelength is

$$\lambda_c = \frac{2\pi r_0}{u'_{nr}} \qquad (6.121)$$

where u'_{nr} is the rth root of $\partial J_n(ur)/\partial r = 0$.

It can be seen from Eq. (6.121) that the size of the focus spot decreases as the terms of the summation increase. This in turn requires a larger lens aperture and a higher light source frequency. However, in practice only a finite limited focus size can be achieved. Thus let there be a finite summation N for Eq. (6.120) and let the source wavelength be smaller as compared with the cutoff wavelength of Eq. (6.121). Then a finite radius of the focus spot can be defined:

$$\Delta r \geq \frac{1.22 r_0}{N} \tag{6.122}$$

The corresponding focal area is

$$\Delta A = 2\pi (\Delta r)^2 \geq \pi \frac{(1.22 r_0)^2}{N^2} \tag{6.123}$$

To avoid any possible reflection of the wave guide, we assume that the wave guide is properly terminated by a matched impedance.

To compute the amount of entropy required for this observation, it is necessary for us to assume that short light pulses of duration Δt are used. To determine the finite number of degrees of freedom, we let the light pulses be repeated periodically at every t_0. If we take the effective bandwidth of the pulse (i.e., the spectrum extended from $v = 0$ up to $1/2\,\Delta t$, instead of $1/\Delta t$), the number of degrees of freedom per period of t_0 is

$$M \geq \frac{t_0}{2\,\Delta t} \tag{6.124}$$

Thus the light beam has a total of $N^2 M$ degrees of freedom per period t_0.

Let us first consider $hv \ll kT$, low-frequency observation. Since at $hv \ll kT$ each degree of freedom has an energy of kT, the total thermal energy per period of t_0 is approximately

$$E_T \simeq N^2 M k T \tag{6.125}$$

To overcome this thermal background, we use energy from a light source greater than this amount [see Eq. (6.45)]:

$$\Delta E \geq N^2 M k T [\ln \alpha + 0.367] \qquad \alpha \gg 1 \tag{6.126}$$

The corresponding entropy increase is

$$\Delta S = \frac{\Delta E}{T} \geq N^2 M k [\ln \alpha + 0.367] \tag{6.127}$$

Let us now determine the amount of information per period of t_0. But in order to do so we must first determine the accuracy of the focal spot:

$$\mathcal{A} = \frac{A}{\Delta A} \cdot \frac{t_0}{\Delta t} \tag{6.128}$$

where $A = \pi r_0^2$, the total field of view. Thus the amount of information obtained is

$$I = \log_2 \mathcal{A} = \log_2 \left[2M \left(\frac{N}{1.22} \right)^2 \right] \tag{6.129}$$

It is also noted from Eq. (6.128) that the entropy of Eq. (6.127) can be written

$$\Delta S \geq \frac{(1.22)^2}{2} \mathcal{A} k [\ln \alpha + 0.367] \qquad \alpha \gg 1 \tag{6.130}$$

Thus we see that

$$\Delta S - I k \ln 2 > 0 \tag{6.131}$$

From Eq. (6.128), we see that for a larger field of view A corresponds to higher accuracy; hence a larger amount of information is obtained. However, from Eq. (6.130), we see that the entropy is even greater than the information obtained. This result is one of the basic facts in practical problems: Observation of a focus without field and time limitations eventually leads to an infinite cost in entropy (energy). This is one of the results that Toraldo di Francia [17] obtained. He has pointed out that a two-point resolution is impossible unless the observer has *a priori* an infinite amount of information about an object. Thus infinite entropy (energy) is required. In all practical problems, observations are made under finite limitations of field and time.

Finally, in the high-frequency case, $h\nu \gg kT$, although high accuracy and high reliability can be obtained, entropy required is even greater than for low-frequency observation. This increase in entropy is essentially what we trade for high accuracy and reliability.

6.6 UNCERTAINTY AND OBSERVATION

We now examine the fundamental constraints of physical observations. As we are all aware, all physical problems are ultimately restricted by certain limitations. When quantum conditions are in use, all limitations are essentially imposed by the basic Heisenberg uncertainty principle and by wave mechanics. We discuss the basic limitation in observation imposed by the uncertainty principle. Time and energy smaller than Planck's constant cannot be observed or measured simultaneously [18]:

$$\Delta E \, \Delta t \geq h \tag{6.132}$$

where ΔE can be regarded as the unknown energy perturbation of the observation system and specimen, and Δt is the observation time accuracy.

In applying the Heisenberg uncertainty principle to our observation problems, let us refer to Sec. 6.1, the statistical analysis of observations made by radiation. That is, we compare the energy ΔE required for an observation with the mean-square thermal fluctuation of the photodetector $\gamma k T$, where γ is the number of degrees of freedom. Essentially, this is the number of low-frequency vibrations ($h\nu \ll kT$). Thus if $\Delta E < \gamma k T$, then according to the uncertainty relation of Eq. (6.132) we have

$$\Delta t \gg \frac{h}{\gamma k T} \tag{6.133}$$

From the above inequality we see that a larger time resolution Δt can be obtained with low-frequency observation. Since ΔE is small, the perturbation due to the time resolution is very small and can by comparison be ignored.

However, if the radiation frequency ν becomes greater, such that $\Delta E = h\nu > \gamma k T$, then using the uncertainty relation of Eq. (6.132) we have

$$\Delta t \leq \frac{h}{\gamma k T} \tag{6.134}$$

In this case we see that, as the radiant energy required for the observation increases, the more accurately the time resolution can be measured and observed. But the perturbation of the observation is also greater. Thus the time resolution Δt obtained with the observation may not even be correct, since ΔE is large.

The observation system is generally assumed to be nonperturbable, as in the classical theory of light, in which case precise observation can always be obtained and can be repeated many times with consistent results. In other words, the observations provide reproducible results and, as we have noted, this assumption is generally true for the many-particles problem and a large number of quanta in observation. With the classical theory the accuracy of observation is not expected to be too great, since this condition is imposed far away from the limitation of the uncertainty principle:

$$\Delta E \, \Delta t \gg h \tag{6.135}$$

or, equivalently,

$$\Delta p \, \Delta x \gg h \tag{6.136}$$

where Δp and Δx are the respective momentum and position errors. From Eqs. (6.135) and (6.136) it can be seen that in order to observe the quantities E, t, p, and x the errors of ΔE, Δt, Δp, and Δx must not be too small.

However, as pointed out earlier, when quantum conditions occur, a nonperturbing system simply does not exist. When the higher-quantum hv is used, a certain perturbation within the system is bound to occur; hence high accuracy in high-frequency observation is limited by the uncertainty principle.

Let us now investigate the problem of observing extremely small distances. In Sec. 6.5 [Eq. (6.101)] we stated that, for the observation of a small distance Δx between two particles, one must use a light source having a wavelength λ that satisfies the condition

$$\lambda \le 2\,\Delta x \tag{6.137}$$

Since Δx is assumed to be extremely small, a high-frequency light source is required for the observation. Thus we see that high-frequency observation corresponds to higher momentum:

$$p = \frac{h}{\lambda} \ge \frac{h}{2\,\Delta x} \tag{6.138}$$

In turn, this high-frequency source of radiation corresponds to a higher-quantum hv, in which it interacts with the observed specimen as well as with the observing equipment, and ultimately affects the whole observing system, causing it to change its momentum which may vary from $-p$ to p. Thus the change in momentum of the observing system is

$$\Delta p = 2p \gg \frac{h}{\Delta x} \tag{6.139}$$

Since the radiant energy provided can be written

$$\Delta E = hv = \frac{hc}{\lambda} \ge \frac{hc}{2\,\Delta x} \tag{6.140}$$

we can conclude that

$$\Delta E\,\Delta x \ge \frac{hc}{2} \tag{6.141}$$

It can be seen from the above equation that, mathematically speaking, there is no lower limit to Δx as long as ΔE is able to increase. But Eq. (6.139) tells us that, as ΔE increases, the perturbation of the system under observation cannot be ignored. Thus in practice, when ΔE reaches a certain quantity, the precise observation of Δx is obstructed and the observation of smaller and smaller objects presents ever-increasing difficulty. If certain sophisticated observation equipment were available for use in counteracting the perturbations, the cost of observation would be even greater. Eventually, the cost of observation would become so great that no one would be able to afford it.

Finally, let it be emphasized that Heisenberg's principle of uncertain observation is restricted to the ensemble point of view. That is, for a special observation, the uncertainty may be violated. However, we have never been able to predict when this observation will occur. Therefore a meaningful answer to the Heisenberg uncertainty principle is present only in the sense of statistical ensemble average.

6.7 REMARKS

It has been shown that the amount of information obtained from a physical observation depends on the logarithm of the accuracy:

$$I = \log_2 \mathcal{A} \tag{6.142}$$

but the cost in entropy is somewhat higher:

$$\Delta S > Ik \ln 2 \tag{6.143}$$

Thus the overall observation system satisfies the second law of thermodynamics, which says

$$\Delta S - Ik \ln 2 > 0 \tag{6.144}$$

The efficiency of observation has been defined as

$$\eta = \frac{Ik \ln 2}{\Delta S} \leq 1 \tag{6.145}$$

We have seen that, for low-frequency observation, the efficiency is somewhat greater than for high-frequency observation. However, for very small distances or small objects, high-frequency observation is better. And although the efficiency is somewhat less, high-frequency observation results in greater accuracy and reliability. However, it is emphasized that, aside from the energy constraints, in high-frequency observation the ultimate limitation is imposed by Heisenberg's uncertainty principle, that is, the smaller the distances or the objects to be observed, the greater the amount of energy required for the observation. Thus perturbation in the observation system cannot be avoided up to certain energy limits. Furthermore, it is clear that direct observation usually provides greater efficiency. However, in practice, it cannot always be obtained, for example, where very small distances are involved.

We also see that every observation requires a certain compensatory increase in entropy. However, in some practical cases the amount of entropy compensation may be negligible, since it is very small when compared with the total entropy of the observation system.

Furthermore, from Eq. (6.144), we see that observation is essentially an irreversible process from the thermodynamic point of view [19].

REFERENCES

1. M. Born and E. Wolf, *Principles of Optics*, 2nd rev. ed., Pergamon, New York, 1964.
2. F. T. S. Yu, *Introduction to Diffraction, Information Processing, and Holography*, MIT Press, Cambridge, Mass., 1973.
3. D. Gabor, "Light and Information," in E. Wolf, ed., *Progress in Optics*, vol. I, North-Holland, Amsterdam, 1961.
4. D. Gabor, "Informationstheorie in der Optik," *Optik*, vol. 39, 86 (1973).
5. L. Brillouin, "The Negentropy Principle of Information," *J. Appl. Phys.*, vol. 24, 1152 (1953).
6. L. Brillouin, *Science and Information Theory*, Academic, New York, 1956.
7. L. Brillouin, *Scientific Uncertainty and Information*, Academic, New York, 1964.
8. L. Szilard, "Über die Entropieverminderung in Einem Thermodynamischen System bei Eingriffen Intelligenter Wesen," *Z. Phys.*, vol. 53, 840 (1929).
9. L. Szilard, "On the Decrease of Entropy in a Thermodynamic System by the Intervention of Intelligent Beings" (translated by A. Rapaport and M. Knoller), *Behav. Sci.*, vol. 9, 301 (1964).
10. F. T. S. Yu, "Observation, Information, and Optical Synthetic Aperture of Spherical Lenses," *Optik*, vol. 38, 425 (1973).
11. R. Fano, *Transmission of Information*, MIT Press, Cambridge, Mass., 1961.
12. M. Schwartz, *Information Transmission, Modulation, and Noise*, 2nd ed., McGraw-Hill, New York, 1970.
13. W. W. Harman, *Principles of the Statistical Theory of Communication*, McGraw-Hill, New York, 1963.
14. I. Selin, *Detection Theory*, Princeton University Press, Princeton, N.J., 1965.
15. J. D. Kraus and K. P. Carver, *Electromagnetics*, 2nd ed., McGraw-Hill, New York, 1973.
16. S. Ramo and J. R. Whinnery, *Fields and Waves in Modern Radio*, John Wiley, New York, 1953.
17. G. Toraldo di Francia, "Resolving Power and Information," *J. Opt. Soc. Am.*, vol. 45, 497 (1955).
18. J. L. Powell and B. Crasemann, *Quantum Mechanics*, Addison-Wesley, Reading, Mass., 1961.
19. F. W. Sears, *Thermodynamics, the Kinetic Theory of Gases, and Statistical Mechanics*, Addison-Wesley, Reading, Mass., 1953.

7
Image Restoration and Information

In Chap. 6 we presented in detail the relationship between observation and information, arriving at a very important result, namely, that for every observation there is a compensatory increase in entropy of other sources. We have shown that every physical observation, as well as every physical device, is ultimately limited by Heisenberg's uncertainty principle from quantum mechanics. In observing the most minute object or a small distance, there is a practical limit at which great accuracy cannot be achieved. Also, the smaller the distances or objects to be observed are, the more costly the energy (or entropy) required. In some instances, the amount of energy required may even approach infinity.

Now we come to the problem of image restoration and information. We first discuss image restoration techniques and certain limitations that are encountered. We illustrate that it is not possible to achieve infinite precision even with finite *a priori* knowledge of the image. We also demonstrate that the restoration of lost information may lead to an infinite amount of energy (or entropy) compensation. Finally, an example of the restoration of a smeared photographic image is given.

7.1 IMAGE RESTORATION

A smeared or out-of-focus photographic image can be restored by means of a coherent optical processor [1, 3]. However, as will be shown, the restored image will not be better than that in a photograph taken with the *maximum allowable exposure time*, that is, the longest possible exposure time during which the recorded image will not smear significantly. The allowable time depends on the size and details of the object being photographed. For example, the smaller the size or the finer the detail, the shorter the allowable time required for exposure. In other words, the restored image theoretically can only approach the case in which the image is recorded without being smeared or becoming out of focus.

In general, the problem of image restoration can be divided into two categories:

1. Restoration from the distorted image
2. Restoration by superimposing the images that were distorted by smearing or by being out of focus

In the first category, it is possible for the image to be restored, but the process does not use the smeared image imposed on the film. In the second category, it is possible not only to restore the image, but also to utilize the excessive recording of the smeared image.

We see that, if spatial random noise is ignored, the results of the above two restorations are essentially identical, since they provide the same amount of information. However, if the random noise of the recorded image is taken into account, then the second restoration category will have a higher information content. We show that the second restoration category is impossible to achieve in practice [4, 5]

For simplicity, a few techniques of image restoration in which the smear is due to linear image motion are discussed. Certain physical constraints of the restoration are also included in the discussion.

Let an image be distorted by some physical means; then the distorted image may be described by its Fourier transform:

$$G(p) = S(p)D(p) \tag{7.1}$$

where $G(p)$ is the distorted image function, $D(p)$ is the distorting function, $S(p)$ is the distortionless image function, and p is the spatial frequency.

We know that the distortionless image functions $S(p)$ can be recovered if we know the distorting function $D(p)$:

$$S(p) = G(p) \cdot \frac{1}{D(p)} \tag{7.2}$$

First, we momentarily disregard the physical realizability of the inverse filter $1/D(p)$. Then we ask whether this *inverse filtering process* increases the information content of the distorted signal $G(p)$. The answer is no, it does not, because we know exactly how it is distorted. The additional amount of information we gained from inverse filtering was essentially derived from the *a priori* knowledge of $D(p)$. One might argue that if this method of recovering the image does not increase the information content, why do we go to the trouble of restoring it? The answer apparently involves an image recognition problem. Since we recognize $G(p)$, as well as $S(p)$, through the knowledge of $D(p)$, these two images are essentially no different from each other—at least from the information standpoint. But for those who do not know how the image was distorted *a priori*, certainly a more realistic image $S(p)$ contains more information than a distorted one. This is the major reason we go to the trouble of restoring the image. However, it is noted that the information gain apparently comes from the compensation of certain energy (entropy) in converting $G(p)$ to $S(p)$.

Let us now assume that a resolvable point image of constant irradiance I is projected on the recording medium of a camera. If the point object moves at a constant velocity v, and the exposure time is t, then the moving-point image recorded on the photographic film will smear into a straight line of length $\Delta x = mvT$, with m a proportionality constant. The corresponding transmittance may be described by the expression

$$f(x) = \begin{cases} A & \text{for } -\frac{\Delta x}{2} \leq x \leq \frac{\Delta x}{2} \\ 0 & \text{otherwise} \end{cases} \tag{7.3}$$

where A is a positive constant proportional to I. The corresponding Fourier transform is

$$F(p) = A \Delta x \frac{\sin(p \Delta x/2)}{p \Delta x/2} \tag{7.4}$$

In the image restoration we apply the inverse filtering process, giving

$$F(p) \frac{A}{F(p)} = A\mathbf{F}[\delta(x)] \tag{7.5}$$

where $\mathbf{F}$ denotes the Fourier transform, and $\delta(x)$ is the Dirac delta function.

From Eq. (7.4) it can be seen that an inverse filter $A/F(p)$ cannot be realized in practice, since it contains poles at $p = 2n\pi/\Delta x$, $n = 0, 1, 2, \ldots$. However, if one assumes that the inverse filter can be completely or partially realized, then the restored image is at best equal to $A\delta(x)$, the minimum resolvable point image. In other words, the inverse filtering process restores the image but does not increase the overall image spectrum over that obtained with the maximum allowable exposure. But is it possible to have a physical filter $H(p)$ such that

$$F(p) \cdot H(p) = B\mathbf{F}[\delta(x)] \tag{7.6}$$

where B is a positive real constant. The question is can B be *greater* than A?

The answer to this question is no, since a coherent optical information processing system is *passive*. In view of the practical complications of filter synthesis, we are led to a different approach to our solution.

Again, suppose a moving image $f(x)$ is recorded on film, and that the length is Δx. Image smearing can be controlled by exposure modulation, such as that composed of a finite sequence of identical functions. Then the transmittance of the recorded function can be described as

$$g(x) = f(x) + f(x - \Delta l) + f(x - 2\Delta l) + \cdots + f(x - N\Delta l)$$
$$= \sum_{n=0}^{N} f(x - n\Delta l) \qquad (7.7)$$

where $N = \Delta x / \Delta l$, and Δl is the incremental translation of $f(x)$. The corresponding Fourier transform is

$$G(p) = \sum_{n=0}^{N} F(p) e^{-ipn\Delta l} \qquad (7.8)$$

where $G(p)$ and $F(p)$ are the Fourier transforms of $g(x)$ and $f(x)$, respectively. From Eq. (7.8), the image restoration may be attempted by

$$G(p)H(p) = F(p) \qquad (7.9)$$

where the filter is

$$H(p) = \frac{1}{\sum_{n=0}^{N} e^{-ipn\Delta l}} \qquad (7.10)$$

Now if we properly translate the x axis of Eq. (7.7), then by the *Lagrangian identity* [6] the denominator of Eq. (7.10) can be written

$$\sum_{-N/2}^{N/2} e^{-inp\Delta l} = \frac{\sin(N/2 + \frac{1}{2})p\,\Delta l}{\sin \frac{1}{2} p\,\Delta l} \qquad (7.11)$$

For a large N and a small Δl, Eq. (7.10) can be approximated by

$$H(p) \simeq \frac{1}{2} \frac{p\Delta x/N}{\sin(p\,\Delta x/2)} \qquad (7.12)$$

Practically speaking, $H(p)$ is not a physically realizable function. However, if we assume $H(p)$ is physically realizable, then by this same interpretation we can say that, at best, the restored image can only be equal to the image recorded with the maximum allowable exposure time.

But because of the physical unrealizability of the filter, we must modify the exposure-modulating process. From Eqs. (7.9) and (7.10) it appears that the filter $H(p)$ can be made physically realizable if unity is added to the denominator of Eq. (7.10):

$$H(p) = \frac{1}{1 + \sum_{n=0}^{N} e^{-ipn\Delta l}} \tag{7.13}$$

We see that, if the filter of Eq. (7.13) is used, the physical realizability condition is

$$|H(p)| \leq 1 \quad \text{for all } p \tag{7.14}$$

One of the obvious procedures in making $H(p)$ physically realizable is to control the exposure properly, so that the resulting transmittance is

$$\begin{aligned} g(x) &= 2f(x) + f(x - \Delta l) + f(x - 2\Delta l) + \cdots + f(x - N\Delta l) \\ &= f(x) + \sum_{n=0}^{N} f(x - n\Delta l) \end{aligned} \tag{7.15}$$

where $N = \Delta x/\Delta l$, and $\Delta x = mvT$. The corresponding Fourier transform is

$$G(p) = F(p)\left(1 + \sum_{n=0}^{N} e^{-ipn\Delta l}\right) \tag{7.16}$$

Thus the inverse filter function is

$$H(p) = \frac{1}{1 + \sum_{n=0}^{N} e^{-inp\Delta l}} \simeq \frac{1}{1 + N\frac{\sin(p\Delta x/2)}{p\Delta x/2}} \tag{7.17}$$

and is a physically realizable filter. However, it can be seen that the restored image spectrum is equal to $F(p)$, which is only one-half of the first image function in Eq. (7.15).

Now we discuss an interesting image restoration technique. If the recorded image can be modulated such that each of the resolvable points can be imaged by itself, then it may be possible to reconstruct the image by means of coherent illumination. Not only will the smearing of the image be corrected, but in addition the overall image spectrum may be improved.

Suppose a movable-point image is recorded on the film by means of a certain shutter process, so that the amplitude transmittance is

$$T(x) = \tfrac{1}{2} - \tfrac{1}{2}\cos\frac{\pi}{\lambda f}x^2 = \tfrac{1}{2} - \tfrac{1}{4}e^{i(\pi/f\lambda)x^2} - \tfrac{1}{4}e^{-i(\pi/f\lambda)x^2} \tag{7.18}$$

where λ is the wavelength of the coherent source, and f is the focal length of

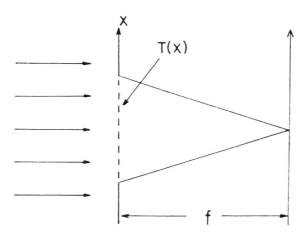

Fig. 7.1 Under coherent illumination.

the one-dimensional *zone lens* [5]. It is noted that a zone lens is similar to a *zone plate* [5], except that the transmittance is a sinusoidal function of x^2.

If this modulated transparency of Eq. (7.18) is illuminated by a monochromatic plane wave front as shown in Fig. 7.1, then the complex light field at the focal length f may be determined by the Fresnel-Kirchhoff theory or by Huygens' principle [7] (Sec. 2.2):

$$U(\alpha) = \int_{-\infty}^{\infty} T(x) u(\alpha - x; k) \, dx \tag{7.19}$$

where

$$u(x) = \frac{1}{f} \exp\left[i \frac{2\pi}{\lambda} \left(f + \frac{x^2}{2f}\right)\right]$$

is known as the *spatial impulse response*.

By substituting Eq. (7.18) into Eq. (7.19), the solution is

$$U(\alpha) = \left[\frac{1}{2}\left(\frac{1}{\lambda}\right)^{1/2} \exp\left(i\frac{\pi}{4}\right) - \frac{1}{4}\left(\frac{1}{2f\lambda}\right)^{1/2} \exp\left[i\frac{\pi}{2f\lambda}\alpha^2 + \frac{\pi}{4}\right]\right. \\ \left. - \frac{1}{4} \exp\left(-i\frac{\pi}{f\lambda}\alpha^2\right) \delta(\alpha) \right. \tag{7.20}$$

where $\delta(\alpha)$ is a Dirac delta function.

It is clear that the first term of Eq. (7.20) is the zero-order diffraction. The second term is the divergent diffraction. The last term corresponds

Image Restoration and Information 171

to the convergent diffraction. In other words, a zone lens such as the one of Eq. (7.18) is equivalent to three lenses (i.e., dc, divergent, and convergent). However, if we examine Eq. (7.20) a bit more carefully, we see that the convergent image term is reduced by a factor of 4 (i.e., a factor of 16 with respect to irradiance) compared with the one used for the maximum allowable exposure time. Obviously, the excessive recording (energy) on the film due to the modulation was not effectively used. On the contrary, this excess recording has been converted into useless diffraction in the zero and divergent terms. Thus from this simple analysis we are led to doubt that there exists a realizable shutter modulation technique in which the excessive recording can be fully or partially converted into useful diffraction. The restoration of a smeared image may be seen as an energy-and-time problem for which a physically realizable solution involving the expenditure of a finite amount of energy does not exist.

In general, a smeared image can be represented in a three-dimensional orthogonal space with a closed time interval ($t_1 \leq t \leq t_2$). The smeared image can therefore be regarded as a time-and-space problem. If we claim to be able to restore the image due to smearing by piling up the excessive recording on the film, it is equivalent to saying it is possible to separate the times and the spaces from the smeared image. Of course, from the basic theory of relativity [8], it is impossible for us to do so without expending an excessive amount of entropy.

7.2 UNCERTAINTY AND IMAGE RESTORATION

In order to demonstrate that restoration of a smeared image is a problem related to Heisenberg's uncertainty principle [9], we start with the basic inequality of a photographic film:

$$It \geq E_0 \quad (7.21)$$

where E_0 is the minimum energy (exposure) per unit area of the film required for an object to be properly recorded, I is the radiance intensity of the minimum resolvable image, and t is the exposure time of the film. In other words, if the image is recorded below the minimum required exposure E_0, it is impossible for it to be resolved. This minimum exposure E_0 can be called the *threshold level* of the photographic film.

To demonstrate that the foregoing inequality is equivalent to Heisenberg's uncertainty relation, one can simply substitute $E_0 = h\nu$ in Eq. (7.21); thus

$$Et \geq h \quad (7.22)$$

where $E = I/\nu$, h is Planck's constant, and ν is the frequency.

We see that these inequalities give the theoretical limit of the film. The inequality of Eq. (7.21) or (7.22) gives the *detectability* (or *recordability*) of the film. If the situation does not satisfy the detectability it will be impossible for the recorded image to be resolved. In the following, we use this inequality to connect the theory of information.

For stationary objects and film, if the signal-to-noise ratio of the image is high, there is no essential difficulty in satisfying the inequality of Eq. (7.21) or (7.22). However, for moving objects, for film, or for both, there exists a maximum allowable exposure time with which the recorded image will not be distorted as a result of smearing:

$$t \leq t_{max} \tag{7.23}$$

where t_{max} is the maximum allowable exposure time. In general, t_{max} is linearly related to the size of the recorded image.

From the inequality of Eq. (7.22) a trade of energy and exposure time is possible; with high irradiance, the film can record smaller objects. But if the irradiance from the object is low, it may be impossible to reduce the maximum allowable exposure time without violating the detectability conditions.

If we are tempted to satisfy the detectability conditions by letting the exposure time become greater than the allowable time, then the recorded image will be further degraded (as a result of smear). Some of the information recorded in the allowable time interval, $0 \leq t \leq t_{max}$, is partially or completely destroyed by this excessive recording. This amount of information loss can also be regarded as an increase in the entropy of the film. Hence a smeared image contains less information than an unsmeared one under the maximum allowable time conditions. This loss of information due to additional recording is also the physical basis of the unrealizability of the inverse filter discussed in Sec. 7.1. In order to bring back the information loss as a result of smearing, the expenditure of an infinite amount of energy is required.

In coherent optical processing a great deal of energy available is from the source. Can it be used? The answer appears to be no, since the energy in a coherent system is used to convey information; it is therefore not convertible into information. For example, suppose a printed page contains a certain amount of information. If it is located in a dark room, the information content has no way of reaching us. We need a certain quantity of light in order to make the information observable. This quantity of light is the cost in energy paid for the information transmission. Moreover, if this page contains a number of misprinted words, and these misprinted words are independent of the correct ones, then a great amount of information (or energy) will be required for us to restore the intended words.

We conclude that the extra energy (or bits of information) we intended for the film to absorb in order to correct the smear actually further degrades the information content of the image. However, to say that it is possible to reconstruct completely the original image from film degraded by smearing or modulation is equivalent to saying that it is possible to make the recording instrument record an unrecordable object after a recording; the contradiction is apparent.

But one can still argue that if an object is embedded in noise (such as turbulence), why is it possible to restore the image by using complex spatial filtering? Before we attempt to answer this question, we should ask whether the source irradiance and the exposure time satisfy the detectability conditions. If they do not, can we restore the imaging object? We cannot. However, if they *do* satisfy the detectability conditions, can we restore the imaging object better than we could in the case without noise? Again, we cannot. At best, we can only approach the case without noise as a limit. Therefore we conclude that the imaging object can be restored (with a probability of error) from the degraded random noise if and only if the intensity from the object and the exposure time of the film fulfill the basic detectability conditions:

$$It \geq E_0 \qquad (7.24)$$

Furthermore, the restored image, at best, can only be equal to the image obtained under the conditions without random noise (such as turbulence).

A few questions can still be raised: Is there a coding process (such as a modulating camera) that can improve the smearing correction? No, because to improve the information transmission through a communication channel, the coding process must take place at the transmitted end—not at the receiving end.

How about coherent detection techniques, such as correlation detection and sampling? Can these improve image restoration? No, since we have no prior knowledge about the object (the recorded image). There is no way to correlate the recorded smeared image. And so we summarize these considerations as follows:

1. A smeared image is basically correctable. However, the corrected results can only approach the case in which the image is recorded with the maximum allowable exposure time ($t = t_{\max}$) of the film. In practice, the corrected result is far below the $t_{\max}$ criterion, because of the unrealizability of the filter.

2. The modulating camera technique, or any other modulating process, is unable to improve the image resolution over that of the maximum allowable time criterion, because the information content of the image is further degraded by this modulating process.
3. The smear correction problem is basically a time-and-space problem. It is physically impossible to restore the image by partially or completely piling up the excessive recording due to smearing.
4. If the irradiance from the object and the exposure time of the film do not satisfy the detectability conditions of the recording medium, it is physically impossible to restore the recorded image. To do so would violate the uncertainty principle. It is possible to restore the recorded image to a point where it equals the image obtained with the maximum allowable exposure time if and only if the detectability conditions are satisfied.
5. The unrealizability of an inverse filter can be explained from the standpoint of entropy and information; the amount of information loss as a result of degradation of the recorded image (such as smearing) can be recovered only with the expenditure of an infinite amount of energy (i.e., infinite entropy compensation).
6. None of the existing coding processes and coherent detection techniques can improve image restoration, because efficient coding must be done at the transmitting end, not at the receiving end. In coherent detection, there is a lack of prior knowledge of the image recorded.

However, some image restoration after smearing is still possible. The optimum image restoration obtainable can be found by means of a mean-square error approximation such as

$$\overline{\varepsilon^2(x, y)} = \lim_{\substack{x_1 \to \infty \\ y_1 \to \infty}} \frac{1}{x_1 y_1} \int_{-x_1/2}^{x_1/2} \int_{-y_1/2}^{y_1/2} [f_0(x, y) - f_d(x, y)]^2 \, dx \, dy$$

(7.25)

where $f_d(x, y)$ is the desired image, $f_0(x, y)$ is the restored image, and the evaluation of the above equation is subject to the physical constraint

$$|H| \leq 1 \qquad (7.26)$$

where H is the transmittance of the correcting filter.

It is emphasized that the restoration of a smeared image has application where the image was recorded without intention of image motion, that is, where a minimum exposure choice could not be made at the time the image was recorded.

7.3 RESOLVING POWER AND INFORMATION

The possibility of resolving power beyond the classical limit of an idealized optical system has been recognized by Coleman [10], Toraldo di Francia [11], Ronchi [12, 13], and Harris [14]. In this section we discuss some of the physical limitations besides the inevitable noise of optical systems (based on Ref. 15).

There are two long-established theorems from analytic function theory that have been found to be useful in studies of resolution beyond the classical limit. The first of these two theorems deals with the finite integral of the Fourier transform of a spatially bounded object. This theorem [16] states that the Fourier transform of a spatially bounded function is analytic throughout the entire domain of the spatial frequency plane.

The second theorem [17] states that, if a function of a complex variable is analytic in the region R, it is possible, from knowledge of the function in an arbitrarily small region within R, to determine the whole function within R by means of analytic continuation. As a corollary to this theorem, it can be stated that, if any two analytic functions have functional values that coincide in an arbitrarily small region in the region of analyticity, then the values of these functions must be equal everywhere throughout their common region of analyticity. This is known as the *identity theorem* or the *uniqueness theorem*.

From these two theorems, it is clear that, if the spatial transfer characteristics (i.e, the spatial frequency and phase response) of an optical system are known, then for a bounded object it is possible to resolve the object with infinite precision by means of *analytic continuation*. Furthermore, from the corollary to the second theorem, the ambiguity of two close objects does not exist, and the resolved object is therefore unique.

Infinite precision of object resolution is, however, a mathematical ideal. In Sec. 7.4 we see that infinitely precise solution is physically unrealizable. In fact, the amount of information gained from extension of the complex spectral density of the object, which is limited by the spatial cutoff frequency of the optical system, must come from proceedings with the analytic continuation. We also show that the degree of precision depends on the accuracy of the functional spectrum expansion. The smaller the region of spectral density known, the greater the effort required for better precision of object resolution.

Let the function $f(z)$ represent the complex spatial frequency spectrum of a bounded object (where $z = x + iy$, and x and y are the spatial frequency coordinates). If $f(z)$ is analytic throughout a given region R, and the functional value is assumed to be given over an arbitrarily small region (or

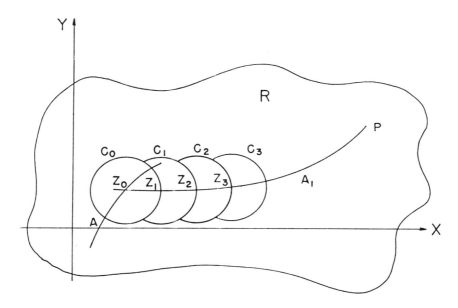

Fig. 7.2 The analytic continuation.

an arc) within R, then by analytic continuation $f(z)$ can be uniquely determined throughout the region R.

Let A denote an arc within R on which the function $f(z)$ is given (Fig. 7.2). Since $f(z)$ is analytic in R, the value of the derivative $f'(z)$ is independent of the manner in which Δz tends to zero. Suppose z_0 denotes a point of A. Then because of the analyticity of $f(z)$, this function can be expanded by Taylor series about z_0, at all points interior to a circle having its center at z_0 and lying within R. Thus

$$f(z) = f(z_0) + \sum_{n=1}^{\infty} \frac{f^n(z_0)}{n!}(z - z_0)^n \qquad |z - z_0| \le r_0 \qquad (7.27)$$

where $f^n(z_0)$ denotes the nth-order derivative of $f(z)$ at z_0, and r_0 is the corresponding radius of the convergent circle within R.

Thus, in order to determine the functional value of $f(z)$ within $|z - z_0| \le r_0$, we must know all the functional derivatives $f^n(z_0)$. For the most general $f(z)$ of course it is not possible to write down the infinite number of derivatives required. If we assume that Eq. (7.27) converges to $f(z)$ with little error when the series is truncated at some finite $n = N$, and also that we can find approximately the associated derivatives $f^n(z_0)$ (which is difficult to achieve in practice; see Ref. 7.18), then the functional value of $f(z)$ within the convergent circle C_0 can be approximated by

$$f(z) \simeq f(z_0) + \sum_{n=1}^{N} \frac{f^n(z_0)}{n!}(z-z_0)^n \qquad (7.28)$$

Now consider any other point P in R. Let A_1 denote a curve connecting z_0 and P and lying in the interior of R (Fig. 7.2), and let the radius r_0 be equal to the shortest distance between the curve A_1 and the boundary R. Let z_1 be a point on curve A_1 within the circle C_0, and let C_1 be a circle with its center at z_1 and radius r_0. Then the function $f(z)$ within C_1 is approximated by

$$f(z) \simeq f(z_1) + \sum_{n=1}^{N} \frac{f^n(z_1)}{n!}(z-z_1)^n \qquad (7.29)$$

By continuing this process, the curve A_1 can be covered by a finite sequence of circles, $C_0, C_1, C_2, \ldots, C_m$, of radius r_0; thus the functional value of $f(z)$ over the whole of R can be determined. However, from the well-known theorem stating that a spatially bounded object cannot be spatially frequency bounded [19], the complete extension of $f(z)$ requires an infinite sequence of circles to cover the entire spatial frequency plane. Therefore this analytic extension of $f(z)$ also requires an infinite amount of effort (energy), and it is not physically realizable.

Equation (7.29) is the analytic continuation of Eq. (7.28). The error of this approximate extension of $f(z)$ increases as the analytic continuation proceeds. For example, a one-dimensional hypothetical spatial frequency spectrum is shown in Fig. 7.3, where p_c is the spatial cutoff frequency of the optical system. If we allow the complex spatial frequency spectrum to be bounded for all practical purposes, then by an analytic continuation similar to the foregoing we can see that the degree of precision of object resolution increases with an increase in spatial frequency bandwidth. It is emphasized that, strictly speaking, a spatially bounded object cannot be spatially frequency bounded [19]. However, this assumption of boundedness is often made in practice. Furthermore, the amount of information gained by analytic continuation of the spatial frequency spectrum clearly comes from this effort, that is, the expenditure of energy, to obtain the derivatives $f^n(z_0)$. We note that determination of $f^n(z_0)$ is generally difficult to achieve in practice.

We have shown in this section that it is not possible to obtain infinite precision of object resolution in practice. However, considering the uniqueness theorem cited earlier, one might wonder why we need to expand the spatial frequency spectrum, since an arbitrarily small region of the spectrum is sufficient to represent an object. Unfortunately, to utilize this

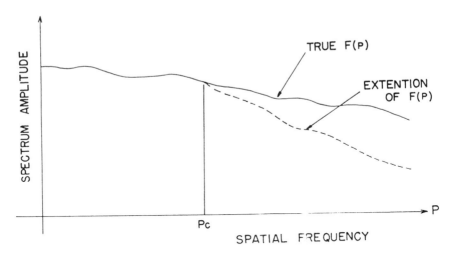

Fig. 7.3 A hypothetical spatial frequency spectrum, showing the extension of the spatial frequency spectrum beyond the diffraction limit of the system.

fact we need to have *a priori* knowledge of objects and their corresponding spatial frequency spectra. If the objects to be detected are finite in number, this may be possible. However, in practice, objects are generally not finite in number; in fact, there may be uncountably many objects. In this case, an infinitely large volume of information storage (i.e., the dictionary of their corresponding spatial frequency spectra) must be available. Such an infinitive capacity is of course physically unrealizable. Incidentally, this result is similar to that obtained by Toraldo di Francia [11]: "A two-point resolution is impossible unless one has *a priori* an *infinite* amount of information about the object."

7.4 COHERENT AND DIGITAL IMAGE ENHANCEMENT

Image enhancement by means of coherent and digital computer techniques has been shown in various applications [20, 21]. We discuss in this section some of their basic distinctions. In a previous section we discussed smeared image restoration by means of an inverse spatial filter. Although such an inverse filter is not physically realizable, there is an approximating technique that may approach it.

A complex spatial filter can be realized by combining an amplitude and a phase filter [5]. Such a phase filter is difficult to synthesize in practice, but these difficulties may be overcome by means of a computer-generated filter [22].

To illustrate the enhancement of a linear smeared photographic image, let us recall Eq. (7.5), where the transfer characteristic of an inverse filter is

$$H(p) = K \frac{\Delta x}{F(p)} \tag{7.30}$$

where

$$F(p) = \Delta x \frac{\sin p(\Delta x/2)}{p(\Delta x/2)}$$

We have seen that $H(p)$ is a bipolar function with a countable number of poles. Such an inverse filter, as pointed out in Sec. 7.1, is not physically realizable. However, if we are willing to accept a certain restoration error, then the inverse complex filter may be written

$$H_1(p) = A(p)\exp[i\phi(p)] \tag{7.31}$$

where $A(p)$ and $\phi(p)$ are the amplitude and phase filter functions, respectively. The physical constraint of such a filter, in coherent optical processing, is of course $0 \leq A(p) \leq 1$. The corresponding restored Fourier spectrum is

$$F_1(p) = F(p)H_1(p) \tag{7.32}$$

Let us define the *relative degree of image enhancement* [23]:

$$\mathcal{D}(A_m) = \frac{\int_{\Delta p} [F_1(p)/\Delta x]\,dp}{A_m \Delta p} \times 100\% \tag{7.33}$$

where A_m denotes the minimum value of the amplitude filter function, and Δp is the spatial bandwidth of interest. Δp is limited by the diffraction limit of the processing system. From Eq. (7.33) we see that a perfect degree of enhancement within Δp is possible, except when A_m approaches zero. But we can also see that the Fourier spectrum also approaches zero as A_m approaches zero. Therefore a perfect degree of enhancement, even within the diffraction limit, is not possible. It is also emphasized that, in coherent optical processing, image enhancement is also limited by the diffraction limit of the imaging system (e.g., the camera). Thus the ultimate limit is restricted by the imaging or the processing system, whichever comes first. Therefore a perfect degree of image enhancement does not apply beyond the diffraction limit.

We see that resolution beyond the diffraction limit can be achieved by means of analytic continuation [14], or by the Karhunen-Loéve expansion

[24]. Both of these techniques are very restricted, as both possess a common practical limitation: expenditure of an infinite amount of energy.

If we accept certain errors, then restoration beyond the diffraction limit can be obtained by a finite extension of the Fourier spectrum, but the amount of information gain, as noted, comes from the compensation of entropy increase in expanding the spectrum. For analytic continuation, it is the computational effort in obtaining the functional derivatives; and for the Karhunen-Loéve expansion, it is the effort in obtaining the eigenfunctions.

Now we come to the distinction between coherent and computer image restoration. In the coherent technique, restoration is basically an analog technique. The image can be processed simultaneously in both spatial and spatial frequency domains. The restoration is, however, limited by the diffraction limits of the imaging and processing systems, whichever comes first.

In computer image restoration the spatial plane is divided into discrete variables. Image processing usually takes place by sequentially sampling the image plane. Thus it is limited by the finite sample points, but not by the diffraction limit of the system. The significant distinction between these techniques is that the coherent system is a *passive* device and the digital system is an *active* one. In coherent spatial filtering, the physical constraint of the restoration filter is

$$|H(p,q)| \leq 1 \tag{7.34}$$

where p and q are the spatial frequency coordinates. However, in the digital filtering technique, it is possible to process the image beyond this constraint.

In concluding this section we point out that it is too early to compare these two image restoration techniques in their present stages. However, if image restoration is meant to be within the diffraction limit, because of the high spatial resolution and relative simplicity in optical handling, then a coherent technique may possess certain advantages. However, if image restoration is to be beyond the diffraction limit, then a digital technique may be more suitable. Our conclusion is that a hybrid system (coherent-digital) may offer a better image restoration technique.

7.5 INFORMATION LEAKAGE THROUGH A PASSIVE CHANNEL

In this section we consider spatial information transmission through a passive optical channel. It is known to optical information specialists that a coherent light field is generally complex. The information-bearing elements

Fig. 7.4 An input-output optical information channel.

can be described by amplitude and phase [5] and are strictly independent variables. Answering the question as to which of these elements is more relevant in certain applications is not the intention of this discussion. However, we treat both amplitude and phase as equally relevant. We determine the information leakage [25] from the basic entropy definitions developed in Chap. 1. The result as applied to image enhancement is also discussed.

The technique of transmitting spatial information through an optical channel is shown in Fig. 7.4. Let us denote the input ensemble of the amplitude and phase signals by $\{A, \phi\}$, and the corresponding output ensemble of the amplitude and phase variables by $\{B, \theta\}$. It is noted that

$$\{A, \phi\} \triangleq A(x, y) \exp[i\phi(x, y)] \tag{7.35}$$

and

$$\{B, \theta\} \triangleq B(\alpha, \beta) \exp[i\theta(\alpha, \beta)] \tag{7.36}$$

where (x, y) and (α, β) are the corresponding input and output spatial coordinate systems. The amount of information the input signal $\{A, \phi\}$ provides, as defined by Shannon [26, 27], is

$$H(A, \phi) = -\int_0^{2\pi}\int_0^{\infty} p(A, \phi)\log_2 p(A, \phi)\, dA\, d\phi \tag{7.37}$$

with

$$\int_0^{2\pi}\int_0^{\infty} p(A, \phi)\, dA\, d\phi = 1 \tag{7.38}$$

where $p(A, \phi)$ is the joint probability density of A and ϕ, with $0 < A < \infty$ and $0 \leq \phi < 2\pi$. The corresponding conditional entropy or *spatial channel equivocation* is

$$H(A, \phi/B, \theta) = -\int_0^{2\pi}\int_0^{\infty}\int_0^{2\pi}\int_0^{\infty} p(A, \phi, B, \theta)\log_2 p(A, \phi/B, \theta)\, dA\, d\phi\, dB\, d\theta$$

with

$$\int_0^{2\pi} \int_0^{\infty} \int_0^{2\pi} \int_0^{\infty} p(A, \phi, B, \theta)\, dA\, d\phi\, dB\, d\theta = 1 \tag{7.40}$$

where $p(A, \phi\ B, \theta)$ is the corresponding input-output joint probability density, and $p(A, \phi/B, \theta)$ is the a posteriori probability density with $0 < B < \infty$ and $0 \le \theta \le 2\pi$. Now the mutual information (as defined in Chap. 1) of the optical spatial channel can be determined:

$$I(A, \phi; B, \theta) = H(A, \phi) - H(A, \phi/B, \theta) \tag{7.41}$$

Similarly, one can write

$$I(A, \phi; B, \theta) = H(B, \theta) - H(B, \theta/A, \phi) \tag{7.42}$$

where

$$H(B, \theta) = -\int_0^{2\pi} \int_0^{\infty} p(B, \theta) \log_2 p(B, \theta)\, dB\, d\theta \tag{7.43}$$

and

$$H(B, \theta/A, \phi) = -\int_0^{2\pi} \int_0^{\infty} \int_0^{2\pi} \int_0^{\infty} p(A, \phi, B, \theta) \log_2 p(B, \theta/A, \phi)\, dA\, d\phi\, dB\, d\theta \tag{7.44}$$

By substitution of Eqs. (7.37) and (7.39) into Eq. (7.42) we have

$$I(A, \phi; B, \theta) = \int_0^{2\pi} \int_0^{\infty} \int_0^{2\pi} \int_0^{\infty} p(A, \phi, B, \theta) \log_2 \frac{p(A, \phi, B, \theta)}{p(A, \phi) p(B, \theta)}\, dA\, d\phi\, dB\, d\theta \tag{7.45}$$

From Eq. (7.45), it is a simple matter to show that

$$I(A, \phi; B, \theta) \ge 0 \tag{7.46}$$

The equality of Eq. (7.46) holds if and only if

$$p(A, \phi, B, \theta) = p(A, \phi) p(B, \theta) \tag{7.47}$$

that is, the input-output ensembles are statistically independent.

From Eq. (7.46) we learn that, on the statistical average, information transmitted through an optical spatial channel is never negative. The extreme case is a complete loss of information transmitted through the channel [$I(A\ \phi; B, \theta) = 0$]. In essence, $I(A, \phi; B, \theta) = 0$ implies that $H(A, \phi/B, \theta) = H(A, \phi)$, the channel equivocation, is equal to the information provided at the input.

With these basic properties of spatial information we are in a position to investigate information leakage through the channel, (i.e., losses during

Image Restoration and Information

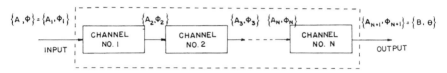

Fig. 7.5 Decomposition of the optical channel into finite cascaded channels.

information transmission). In order to do so, we assume that the optical channel can be decomposed into finite cascaded channels [28, 29] as shown in Fig. 7.5. By $\{A_n, \phi_n\}$ we denote the input signal ensemble applied to the nth cascaded channel, and by $\{A_{n=1}, \phi_{n=1}\}$ the corresponding output signal ensemble, where $n1, 2, \ldots, N$.

It can be seen that there exists a certain relationship between the input and output ensembles at each of the cascaded channels. The overall output signal ensemble (i.e., the set of $\{A_{N+1}, \phi_{N+1}\} = \{B, \theta\}$) therefore depends on the input of $\{A_1, \phi_1\} = \{A, \phi\}$. This dependent property may be described by the conditional probability densities

$$p(A_{N+1}, \phi_{N+1}/A_N, \phi_N, \ldots, A_1, \phi_1) = P(A_{N+1}, \phi_{N+1}/A_N, \phi_N) \quad (7.48)$$

Similarly, by Bayes' theorem [30], a reversal relationship for these cascaded channels can be written

$$p(A_1, \phi_1/A_2, \phi_2, \ldots, A_{N+1}, \phi_{N+1}) = p(A_1, \phi_1/A_2, \phi_2) \quad (7.49)$$

It is noted that the relationship of Eqs. (7.48) and (7.49) is correct only for *linear* cascaded channels.

As the optical information propagates through the cascaded channels, the corresponding entropy equivocations increase:

$$H(A_1, \phi_1/A_{N+1}, \phi_{N+1}) \geq H(A_1, \phi_1/A_N, \phi_N) \geq \cdots \geq H(A_1, \phi_1/A_2, \phi_2) \quad (7.50)$$

Now let us investigate the termwise relationship of these entropy equivocations. We have, for example

$$H(A_1, \phi_1/A_3, \phi_3) - H(A_1, \phi_1/A_2, \phi_2)$$
$$= -\int_0^{2\pi}\int_0^{\infty}\int_0^{2\pi}\int_0^{\infty} p(A_1, \phi_1, A_3, \phi_3) \log_2 p(A_1, \phi_1/A_3, \phi_3) \, dA_1 \, d\phi_1 \, dA_3 \, d\phi_3$$
$$+ \int_0^{2\pi}\int_0^{\infty}\int_0^{2\pi}\int_0^{\infty} p(A_1, \phi_1, A_2, \phi_2) \log_2 p(A_1, \phi_1/A_2, \phi_2) \, dA_1, d\phi_1 \, dA_2 \, d\phi_2$$

$$(7.51)$$

Equation (7.51) can be written

$$H(A_1, \phi_1/A_3, \phi_3) - H(A_1, \phi_1/A_2, \phi_2)$$
$$= \iiiint_{\substack{\{A_2,\phi_2\} \\ \{A_3,\phi_3\}}} p(A_2, \phi_2, A_3, \phi_3) \iint_{\{A_1,\phi_1\}} p(A_1, \phi_1/A_2, \phi_2, A_3, \phi_3) \qquad (7.52)$$
$$\cdot \log_2 \frac{p(A_1, \phi_1/A_2, \phi_2, A_3, \phi_3)}{p(A_1, \phi_1/A_3, \phi_3)} dA_1 \, d\phi_1 \, dA_2 \, d\phi_2 \, dA_3 \, d\phi_3$$

Then by the well-known fact [31]

$$-\int_{-\infty}^{\infty} p(x) \log \frac{p(x)}{g(x)} \, dx \geq 0 \qquad (7.53)$$

where

$$\int_{-\infty}^{\infty} p(x) \, dx = \int_{-\infty}^{\infty} g(x) \, dx = 1$$

where $p(x)$ and $g(x)$ are probability density functions in Eq. (7.52), it can be shown that

$$H(A_1, \phi_1/A_3, \phi_3) - H(A_1, \phi_1/A_2, \phi_2) \geq 0 \qquad (7.54)$$

or, equivalently,

$$H(A_1, \phi_1/A_3, \phi_3) \geq H(A_1, \phi_2/A_2, \phi_2) \geq 0 \qquad (7.55)$$

The result of Eq. (7.54) indicates that the entropy equivocation increases as the information flows through the linear cascaded channels. Then by a simple induction principle one can show that

$$H(A_1, \phi_1/A_{N+1}, \phi_{N+1}) \geq H(A_1, \phi_1/A_n, \phi_n) \qquad \text{for } n = 2, 3, \ldots, N$$
$$(7.56)$$

From Eq. (7.56), it can be shown that the corresponding mutual information of the cascaded channels is

$$I(A_1, \phi_1; A_2, \phi_2) \geq I(A_1, \phi_1; A_3, \phi_3) \geq \cdots \geq I(A_1, \phi_1; A_{N+1}, \phi_{N+1})$$
$$(7.57)$$

This indicates that a passive optical channel has a tendency to leak information.

The useful result of Eq. (7.57) as applied to electrical communication was first recognized by Woodward [32] and, although it may seem trivial, it should not be overlooked in the application of optical information processing. It is also interesting to note that the equalities in Eq. (7.57) hold if and only if

$$p(A_1, \phi_1/A_2, \phi_2, \ldots, A_{N+1}, \phi_{N+1}) = p(A_1, \phi_1/A_{N+1}, \phi_{N+1}) \qquad (7.58)$$

Equation (7.58) implies that

$$p(A_1, \phi_1/A_2, \phi_2) = p(A_1, \phi_1/A_3, \phi_3) = \cdots = p(A_1, \phi_1/A_{N+1}, \phi_{N+1}) \qquad (7.59)$$

The conditions of Eq. (7.59) imply that the cascaded channel is noiseless and undistorted. However, under certain circumstances, despite a somewhat noisy cascaded channel, the equality in mutual information of Eq. (7.57) may hold. But the basic significance of our results is that the information transmitted through a passive optical channel can never be increased. We have seen that the output information, at best, is equal to the input. However, for an active channel, the output information can be made greater than the input, but only through the expenditure of external energy (entropy).

Now we come to the problem of image enhancement. We noted in Sec. 7.4 that a coherent optical channel is passive. The basic constraints are primarily the diffraction limit and spatial filter synthesis:

$$|H(p, q)| \leq 1 \qquad (7.60)$$

Thus we emphasize that a passive optical channel is not able to enhance the image beyond the information limit of the input. But it is also noted that an optical channel is by no means inferior. For example, in many optical information processings, the information obtained at the output end usually depends on certain observations, in most cases, by photodetectors or the human eye. Thus certain information may exist, but beyond the resolution limit of the detectors. However, if it goes through certain coherent processings, this information may be made perceivable. But again, the output information does not increase beyond the input. In other words, if the information does not exist, then it will be impossible to extract it by means of a passive optical channel. However, if an active channel is employed, then in principle it is possible to enhance the information beyond the diffraction limit. The most interesting techniques for information enhancement beyond the diffraction limit, as noted, are analytic continuation and the Karhunen-Loéve expansion. However, with both these techniques the required expenditure of external energy in some cases can be enormous.

7.6 RESTORATION OF BLURRED IMAGES

The restoration of a smeared photographic image by means of coherent optical spatial filtering has been briefly mentioned elsewhere. Some of the physical constraints discovered from the standpoint of information theory have also been discussed in preceding sections. The synthesis of a complex spatial filter to reduce the effect of blurring is now discussed.

As noted, a complex spatial filter may be realized by combining an amplitude filter and a thin-film phase filter. Such a synthesis may be accomplished by means of a holographic technique. The preparation of such a phase filter has been studied by Stroke and Zech [33] for the restoration of blurred images, and by Lohmann and Paris [22] for optical data processing.

In this section we consider the synthesis of a filter that, when combined with an amplitude filter, can be used for restoration of an image that has been blurred. The complex filtering process discussed may be able to correct some blurred images, but the result is by no means an optimum filtering.

We recall that the Fourier transform form of a linearly distorted (blurred) image is [Eq. (7.1)]

$$G(p) = S(p)D(p) \tag{7.61}$$

where $G(p)$ is the distorted image function, $S(p)$ is the undistorted image function, $D(p)$ is the distorting function of the imaging system, and p is the spatial frequency. Then the corresponding inverse filter transfer function for the restoration is

$$H(p) = \frac{1}{D(p)} \tag{7.62}$$

As noted previously, the inverse filter function is generally not physically realizable, particularly for blurred images due to linear motion or defocusing. If we are willing, however, to accept a certain degree of error, then an approximate inverse filter may be obtained. For example, let the transmission function of a linear smeared point image [Eq. (7.3)] be

$$f(\xi) = \begin{cases} 1 & \text{for } -\frac{\Delta\xi}{2} \leq \xi \leq \frac{\Delta\xi}{2} \\ 0 & \text{otherwise} \end{cases} \tag{7.63}$$

where $\Delta\xi$ is the smear length. If a transparency satisfying Eq. (7.63) is inserted in the input plane P_1 of a coherent optical data processor, as shown in Fig. 7.6, the resultant complex light field on the spatial frequency plane will be

$$F(p) = \Delta\xi \frac{\sin(p\Delta\xi/2)}{p\Delta\xi/2} \tag{7.64}$$

Image Restoration and Information

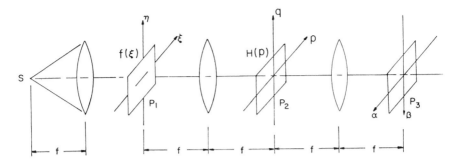

Fig. 7.6 A coherent optical data processor.

which is essentially the Fourier transform of the smeared-point image. A plot of the Fourier spectrum given by Eq. (7.64) is shown in Fig. 7.7. It can be seen that the Fourier spectrum is bipolar.

In principle, this smeared image can be corrected by means of inverse filtering. Thus a suitable inverse filter function is

$$H(p) = \frac{p\,\Delta\xi/2}{\sin\left(\dfrac{p\,\Delta\xi}{2}\right)} \tag{7.65}$$

The transfer characteristic associated with Eq. (7.65) is given in Fig. 7.8. The inverse filter function itself is not only a bipolar but also an infinite-poles function. Thus the filter is not physically realizable. However, if we are willing to sacrifice some of the resolution, then an approximate filter function may be realized. In order to do so, we combine the amplitude filter of Fig. 7.8 with the independent phase filter of Fig. 7.9. The transfer function of this combination is

$$H(p) = A(p)e^{i\phi(p)} \tag{7.66}$$

If this approximated inverse filter is inserted in the spatial frequency plane of the data processor of Fig. 7.6, the restored Fourier transfer function will be

$$F_1(p) = F(p)H(p) \tag{7.67}$$

If we let T_m be the minimum transmittance of the amplitude filter, then the restored Fourier spectrum of the point image is that shaded spectrum

188   Chapter 7

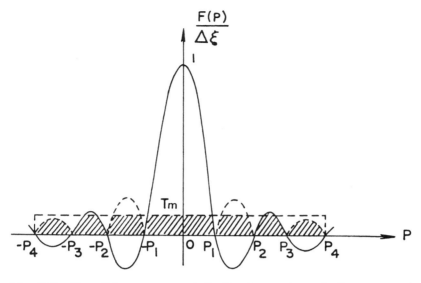

Fig. 7.7 The solid curve represents the Fourier spectrum of a linear smeared point image. The shaded area represents the corresponding restored Fourier spectrum.

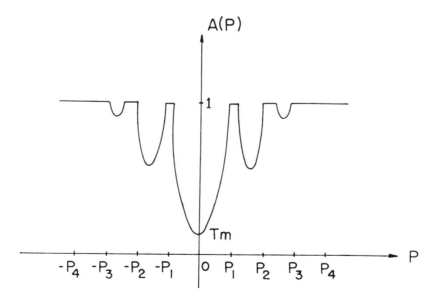

Fig. 7.8 Amplitude filter.

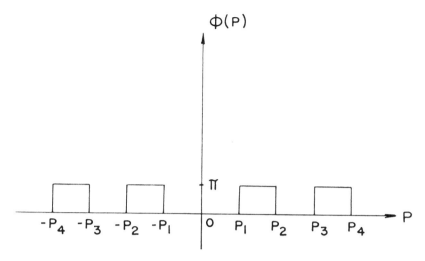

Fig. 7.9 Phase filter.

shown in Fig. 7.7. We define the relative degree of image restoration [Eq. (7.33)] as

$$\mathcal{D}(T_m) = \frac{\int_{\Delta p} [F(p)H(p)/\Delta \xi]/dp}{T_m \Delta p} \times 100\% \qquad (7.68)$$

where Δp is the spatial bandwidth of interest. In Fig. 7.7, for example, $\Delta p = 2p_4$. From Eq. (7.68) we can plot the degree of image restoration as a function of T_m (Fig. 7.10). We see that a perfect restoration is approached as T_m approaches zero. However, at the same time the restored Fourier spectrum is also vanishing, and no image can be reconstructed. Thus perfect restoration cannot be achieved in practice. These considerations aside, it seems that noise (caused by film granularity and speckling) is the major limiting factor in image restoration. In order to achieve a high degree of restoration, it appears that a lower transmittance T_m is required, and the restored Fourier spectrum will therefore be weaker. In turn, a lower signal-to-noise ratio of the restored image will result. Therefore, when considering the noise problem, it is clear that an optimum value T_m must be obtained, at least in practice, for optimum image restoration.

As mentioned earlier in this section, a phase filter can be synthesized by means of a holographic technique (perhaps by a computer-generated

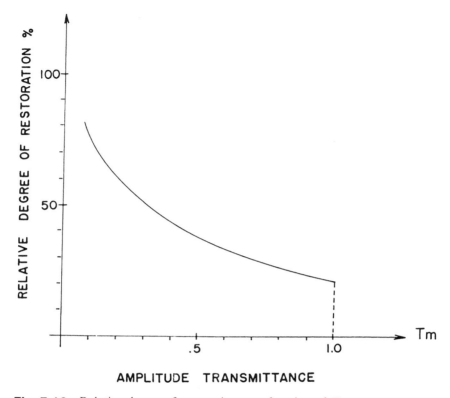

Fig. 7.10 Relative degree of restoration as a function of T_m.

hologram). We now see how such a phase filter works in image restoration. Let us assume that the transmittance of a holographic phase filter is

$$T(p) = \tfrac{1}{2}\{1 + \cos[\phi(p) + \alpha_0 p]\} \tag{7.69}$$

where α_0 is an arbitrarily chosen constant, and

$$\phi(p) = \begin{cases} \pi & \text{for } p_n \leq p \leq p_{n+1}, n = \pm 1, \pm 3, \pm 5, \ldots \\ 0 & \text{otherwise} \end{cases} \tag{7.70}$$

With the amplitude filter added, the complex filter function can be written

$$H_1(p) = A(p)T(p) = \tfrac{1}{2}A(p) + \tfrac{1}{4}[H(p)e^{i\alpha_0 p} + H^*(p)e^{-i\alpha_0 p}] \tag{7.71}$$

where $H(p)$ is the approximate inverse filter function. Note also that $H(p) = H^*(p)$, because of Eq. (7.70).

Image Restoration and Information

Now if this complex filter $H_1(p)$ is inserted in the spatial frequency plane of Fig. 7.6, then the complex light field behind P_2 will be

$$F_2(p) = \tfrac{1}{2} F(p)A(p) + \tfrac{1}{4}[F(p)H(p)e^{i\alpha_0 p} + F(p)H^*(p)e^{-i\alpha_0 p}] \qquad (7.72)$$

We see that the first term of Eq. (7.72) is the restored Fourier spectrum due to the amplitude filter alone, which is diffracted onto the optical axis at

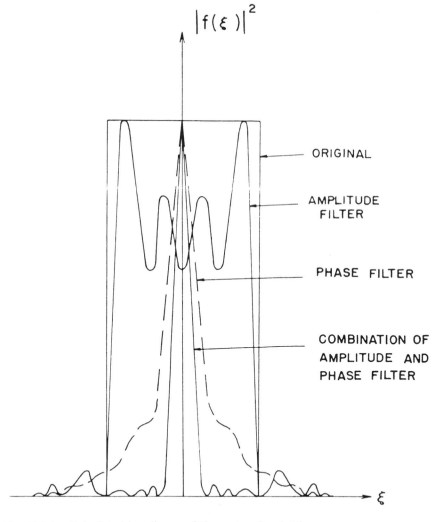

Fig. 7.11 Calculated irradiance of the restored point image.

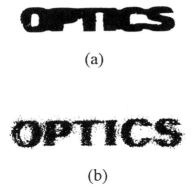

Fig. 7.12 Image restoration with optical processing. (a) Blurred image, (b) restored image.

the output plane P_3 of Fig. 7.6. The second and third terms are the restored Fourier spectra of the smeared image; the restored images due to these terms are diffracted away from the optical axis at the output plane and centered at $\alpha = \alpha_0$ and $\alpha = -\alpha_0$, respectively. As an illustration, Fig. 7.11 shows the calculated irradiance of a restored point image blurred by linear motion; obtained with an amplitude filter, a phase filter, and a combination of both. In view of these results, we see that the phase restoration filter carries substantial weight for the restoration. This phenomenon is rather consistent with the holographic and the matched filtering processes [34–36], for which the phase distribution is the major information bearing elements that responsible for the image reconstruction and detection [37]. To confirm our claim, an experimental result using such a restoration filter is presented in Fig. 7.12. We see that blurred images can indeed be restored with an optical processor [38].

It is emphasized that the relative degree of restoration is with regard to the spatial bandwidth of interest. It is clear that the ultimate limit of Δp is restricted by the diffraction limits of the optical imaging and processing systems, whichever comes first [39]. Therefore, it does not follow that a high degree of restoration implies restoration beyond the diffraction limit.

REFERENCES

1. J. Tsujiuchi, "Correction of Optical Images by Compensation of Aberrations and Spatial Frequency Filtering," in E. Wolf, ed., *Progress in Optics*, vol. II, North-Holland, Amsterdam, 1963.

2. S. L. Zhuang, T. H. Chao, and F. T. S. Yu, "Smeared Photographic Image Deblurring Utilizing White-Light Processing Technique," *Opt. Lett.*, vol. 6, 102 (1981).
3. G. W. Stroke, F. Furrer, and D. R. Lamberty, "Deblurring of Motion-Blurred Photographs Using Extended-Range Holographic Fourier-Transform Division," *Opt. Commun.*, vol. 1, 141 (1969).
4. F. T. S. Yu, "Image Restoration, Uncertainty, and Information," *J. Opt. Soc. Am.*, vol. 58, 742 (1968); *Appl. Opt.*, vol. 8, 53 (1969).
5. F. T. S. Yu, *Introduction to Diffraction, Information Processing, and Holography*, MIT Press, Cambridge, Mass., 1973.
6. R. V. Churchill, *Fourier Series and Boundary Value Problems*, McGraw-Hill, New York, 1941.
7. M. Born and E. Wolf, *Principle of Optics*, 2nd ed, Pergamon, New York, 1964.
8. J. L. Synge, *Relativity*, North-Holland, Amsterdam, 1955.
9. J. L. Powell and B. Crasemann, *Quantum Mechanics*, Addison-Wesley, Reading, Mass., 1961.
10. H. S. Coleman and M. F. Coleman, "Theoretical Resolution Angles for Point and Line Test Objects in the Presence of a Luminous Background," *J. Opt. Soc. Am.*, vol. 37 572 (1947).
11. G. Toraldo di Francia, "Resolving Power and Information," *J. Opt. Soc. Am.*, vol. 45, 497 (1955).
12. V. Ranchi, Optics, *The Science of Vision*, New York University Press, New York, 1957.
13. V. Ranchi, "Resolving Power of Calculated and Detected Images," *J. Opt. Soc. Am.*, vol. 51, 458 (1961).
14. J. L. Harris, "Diffraction and Resolving Power," *J. Opt. Soc. Am.*, vol. 54, 931 (1964).
15. F. T. S. Yu, "Optical Resolving Power and Physical Realizability," *J. Opt. Soc. Am.*, vol. 59, 497 (1969); *Opt. Commun.*, vol. 1, 319 (1970).
16. E. T. Whittaker and G. N. Watson, *A Course of Modern Analysis*, 4th ed, Cambridge University Press, Cambridge, Mass., 1940.
17. E. A. Guilleman, *The Mathematics of Circuit Analysis*, John Wiley, New York, 1951.
18. J. F. Steffesen, *Interpolation*, Chelsea Publishing, New York, 1950.
19. R. E. A. Paley and N. Wiener, "Fourier Transform in the Complex Domain," *Am. Math. Soc. Colloq.*, vol. 19, 16 (1934).
20. G. W. Stroke, "Optical Computing," *IEEE Spectrum*, vol. 9, 64 (1972).
21. H. C. Andrews, *Computer Techniques in Image Processing*, Academic, New York, 1970.
22. A. W. Lohmann and D. P. Paris, "Computer Generated Spatial Filters for Coherent Optical Data Processing," *Appl. Opt.*, vol. 7, 651 (1968).
23. J. Tsujiuchi, T. Honda, and T. Fukaya, "Restoration of Blurred Photographic Images by Holography," *Opt. Commun.*, vol. 1, 379 (1970).
24. B. R. Frieden, "Evaluation, Design and Extrapolation Methods for Optical

Signals, Based on Use of the Prolate Functions," in E. Wolf, ed., *Progress in Optics*, vol. IX, North-Holland, Amsterdam, 1971.
25. F. T. S. Yu and A. Tai, "Information Leakage through a Passive Optical Channel," *J. Opt. Soc. Am.*, vol. 64, 560 (1974); *Opt. Commun.*, vol. 14, 51 (1975).
26. C. E. Shannon, "A Mathematical Theory of Communication," *Bell Syst. Tech. J.*, vol. 27, 379–423, 623–656 (1948).
27. C. E. Shannon and W. Weaver, *The Mathematical Theory of Communication*, University of Illinois Press, Urbana, 1949.
28. R. A. Silverman, "On Binary Channels and Their Cascades," *IRE Trans. Inf. Theory*, vol. IT-1, 19–27 (1955).
29. N. Abramson, *Information Theory and Coding*, McGraw-Hill, New York, 1963, p. 113.
30. E. Parzen, *Modern Probability Theory and its Applications*, John Wiley, New York, 1960.
31. R. M. Fano, *Transmission of Information*, MIT Press, Cambridge, Mass., 1961.
32. P. M. Woodward, *Probability and Information Theory, with Applications to Radar*, Pergamon, New York, 1953.
33. G. W. Stroke and R. G. Zech, "A Posteriori Image-Correcting *Deconvolution* by Holographic Fourier-Transform Division," *Phys. Lett.*, vol. 25A, 89 (1967).
34. F. T. S. Yu, "Optimal Linearization in Holography," *Appl. Opt.*, vol. 8, 2483 (1969).
35. F. T. S. Yu, "Linear Optimization in the Synthesis of Nonlinear Spatial Filter," *IEEE Trans. Inform. Theory*, vol. IT-17, 524 (1971).
36. F. T. S. Yu and S. Jutamulia, *Optical Signal Processing, Computing, and Neural Networks*, Wiley-Interscience, New York, 1992, sec. 6.
37. F. T. S. Yu, "Exploitation of Incoherent Source for Partially Coherent Processing," *Opt. Eng.*, vol. 36, 1458 (1997).
38. F. T. S. Yu, "Coherent and Digital Image Enhancement, Their Basic Differences and Constraints," *Opt. Commun.*, vol. 3, 440 (1971).

8
Quantum Effect on Information Transmission

Definitions of the entropy theory of information and of channel capacity were presented in previous chapters. It is noted that, for discrete sources and channels, these definitions have a mathematical consistency, which is quite satisfactory from an intuitive point of view. However, a formalistic extension of the theory to a continuous channel in the optical frequency range leads to erroneous results. For instance, the capacity of a continuous additive Gaussian channel is [1] [Eq. (1.139)]

$$C = \Delta v \log_2\left(1 + \frac{S}{N}\right) \quad \text{bits/sec} \tag{8.1}$$

where S is the average signal power N is the average power of a white Gaussian noise, and Δv is the bandwidth of the channel.

We see that, if the average noise power approaches zero, then the channel capacity approaches infinity. This is obviously contradictory to the basic physical constraints. Therefore, as the information transmission moves to high-frequency space, where the quantum effect takes place, the communication channel naturally leads to a discrete model. This is the way the quantum theory of radiation, which replaces the classical wave theory, explains the physical phenomena.

We have in previous chapters on several occasions called attention to the quantum phenomenon of information. However, a precise discussion of the quantum effect on a communication channel has not been treated. Although it was in 1929 that Szilard [2] described a basic concept of information from the statistical thermodynamics point of view, the concept was not actually pursued until 1948, by Shannon [3]. As noted, the work by Shannon was originally based on a pure mathematical standpoint and seemed to have nothing to do with the physical sciences. However, the relationship between information theory and physical science was quickly recognized by Gabor [4] in 1950, only a few years later. He advocated that information theory be treated as a branch of physics. But it was the work of Brillouin [5, 9] in the earlier 1950s that successfully established a profound relationship between physical entropy and information. Since then work on the entropy principle of information has been presented in a broad range of applications. However, it was Stern's [10, 11] work in 1960 that actually explained the quantum effect on a communication channel. Stern's work essentially used Gabor's time-frequency cells (logons) in a technique for approaching the maximum entropy of a photon source. In the absence of noise he obtained several interesting results. Following this trend, in 1962 Gordon [12] successfully derived an expression of information capacity for a narrow-band channel by applying frequency representation of a one-dimensional electromagnetic signal and noise. He also discussed the information capacity of coherent amplification and heterodyne and homodyne receivers. In a paper by Lebedev and Levitin [13] information transmission by radiation through wide-band and narrow-band photon channels was again discussed. the effect of quantum statistics on the communication channel and the paradox of infinite channel capacity in the absence of noise were once again explained. Since Lebedev and Levitin's work closely followed the entropy theory of information, we adopt their approach closely in this chapter.

8.1 PROBLEM FORMULATION AND ENTROPY CONSIDERATION

We consider a communication system consisting of an information source of electromagnetic radiation. It is assumed that the information source radiates an electromagnetic signal with an average power S, and the signal propagated through the electromagnetic channel is assumed to be perturbed by additive thermal noise at temperature T. For simplicity, we assume that the electromagnetic signal is one-dimensional (i.e., the photon fluctuation is restricted to only one polarized state), in which the propagation occurs in the same direction of the wave vectors. It is emphasized that the corresponding occupation number of the quantum levels can be fully described. Obviously, these occupation quantum levels correspond to the *microsignal* structure of the information source. It is therefore assumed that these occupation numbers can be uniquely determined for those representing an input signal ensemble. We further assume that an ideal receiver (an ideal photon counter) is used at the output end of the channel. That is, the receiver is able to detect the specified frequencies of the electromagnetic signal. It is also emphasized that in practice the interaction of the electromagnetic signal and the detector are strictly statistical, that is, a certain amount of information loss is expected at the output end of the receiver. The idealized model of the receiver we have proposed mainly simplifies the method by which the quantum effect on the communication channel can be easily calculated.

In the course of the investigation, we use the entropy principle to calculate the amount of information provided at the input end of the channel. We consider an input signal ensemble $\{a_i\}$, $i = 1, 2, \ldots, n$. Let $P(a_i)$ be the corresponding signal probability. If the input signal is applied to a physical system, then each $\{a_i\}$ is able to bring the system to a state of b_j, where the set $\{b_j\}$, $j = 1, 2, \ldots, n$ represents the *macroscopic* states of the system and each macroscopic state b_j is an ensemble of various *microscopic* states within the system. Let us denote by $P(b_j/a_i)$ the corresponding transitional probability. Then for each applied signal a_i the corresponding conditional entropy $H(B/a_i)$ is

$$H(B/a_i) = -\sum_{j=1}^{n} P(b_j/a_i) \log_2 P(b_j/a_i) \tag{8.2}$$

Thus the entropy equivocation (the average conditional entropy) is

$$H(B/A) = \sum_{i=1}^{n} P(a_i) H(B/a_i) \tag{8.3}$$

Since

$$P(b_j) = \sum_{j=1}^{n} P(a_i) P(b_j/a_i) \tag{8.4}$$

the output entropy is

$$H(B) = -\sum_{j=1}^{n} P(b_j) \log P(b_j) \tag{8.5}$$

The corresponding mutual information provided by the physical system (treated as a communication channel) is

$$I(A; B) = H(B) - H(B/A) \tag{8.6}$$

We see that the corresponding thermodynamic entropy is

$$\Delta S = I(A; B) k \ln 2 \tag{8.7}$$

The entropy concept of Eq. (8.7) can be interpreted as indicating that the microstate of the system acts as a communication channel in the transmission of information. However, it is noted that the result of Eq. (8.6) is derived from a strictly stationary *ergodic* property. In practice, we see that the output signal should be time limited, and the system would have some memory. Thus we conclude that, based on the second law of thermodynamics, the physical entropy of the system (the channel) should be greater than that of the information transfer:

$$\Delta S > I(A; B)k \ln 2 \tag{8.8}$$

Thus, as noted in a previous chapter, the inequality of Eq. (8.8) provides us with the basic connection between physical entropy and information entropy. This is the fundamental relationship between statistical physics and information theory. We use this entropy consideration in the following sections to discuss the quantum effect on a communication channel.

8.2 CAPACITY OF A PHOTON CHANNEL

We can now calculate the information capacity of a photon channel. For simplicity, we consider only the case of additive noise, that is, the case in which the noise statistic within the communication channel does not alter in the presence of a signal. At a specific frequency v, let us denote the mean quantum number of an electromagnetic signal by $\bar{m}(v)$, and the mean quantum number of a noise by $\bar{n}(v)$. Thus the signal plus noise is

$$\bar{f}(v) = \bar{m}(v) + \bar{n}(v) \tag{8.9}$$

Since the photon density (i.e., the mean number of photons per unit time per frequency) is the mean quantum number, the corresponding signal energy density per unit time is

$$E_s(v) = \bar{m}(v) h v \tag{8.10}$$

where h is Planck's constant. Similarly, the noise energy density per unit time is

$$E_N(v) = \bar{n}(v) h v \tag{8.11}$$

Since the mean quantum number of the noise (the blackbody radiation at temperature T) follows Planck's distribution [Eq. (6.9)],

$$E_N(v) = \frac{hv}{\exp(hv/kT) - 1} \tag{8.12}$$

the corresponding noise energy per unit time (noise power) can be calculated as

$$N = \int_\epsilon^\infty E_N(v) dv = \int_\epsilon^\infty \frac{hv}{\exp(hv/kT) - 1} dv = \frac{(\pi kT)^2}{6h} \tag{8.13}$$

where k is a Boltzmann's constant, and the lower integral limit ϵ is an arbitrarily small positive constant. [*Note* : Equation (8.12) is also known as *Bose-Einstein distribution*.]

We see that the minimum amount of entropy transfer required by the signal radiation is

$$\Delta S = \int_0^N \frac{dE_s(T')}{T'} = \int_0^T \frac{dE_s(T')}{dT'} \frac{dT'}{T'} \tag{8.14}$$

where $E_s(T)$ is the corresponding signal energy density per unit time as a function of temperature T', and T is the temperature of the blackbody radiation. Thus in the presence of a signal the output radiation energy per unit time (the power) can be written

$$P = S + N \tag{8.15}$$

where S and N are the signal and the noise power, respectively. Since the signal is assumed to be deterministic (i.e., the microstates of the signal are fully determined), the signal entropy can be considered zero. It is emphasized that the validity of this assumption is mainly based on the independent statistical nature of the signal and the noise, for which the photon statistics follow Bose-Einstein distribution. However, it is also noted, the Bose-Einstein distribution cannot be used for the case of *Fermions*, because, owing to the *Pauli exclusion principle*, the microstates of the noise are restricted by the occupational states of the signal, or vice versa. In other words, in the case of Fermions, the signal and the noise can *never* be assumed to be statistically independent. For the case of Bose-Einstein statistics, we see that the amount of entropy transfer by radiation remains unchanged:

$$H(B/A) = \frac{\Delta S}{k \ln 2} \tag{8.16}$$

Since the mutual information (information transfer) is $I(A; B) = H(B) - H(B/A)$, we see that $I(A; B)$ reaches its maximum when $H(B)$ is a maximum. Thus for maximum information transfer, the signal should be chosen randomly. But the maximum value of entropy $H(B)$ occurs when the ensemble of the microstates of the total radiation (the ensemble B) corresponds to Gibbs' distribution, which corresponds to the thermal equilibrium. Thus the corresponding mean occupational quantum number of the total radiation also follows Bose-Einstein distribution at a given temperature $T_e \geq T$:

$$\bar{f}(v) = \frac{hv}{\exp(hv/kT_e) - 1} \tag{8.17}$$

where T_e can be defined as the effective temperature.

Thus the total radiation entropy (the output entropy) can be determined by the equation

$$H(B) = \frac{1}{k \ln 2} \int_0^{T_e} \frac{dE_N(T')}{dT'} \frac{dT'}{T'} \qquad (8.18)$$

Then in accordance with the definition of channel capacity, as given in Chap. 1, we have

$$C = H(B) - H(B/A) = \frac{1}{k \ln 2} \int_T^{T_e} \frac{dE_N(T')}{dT'} \frac{dT'}{T'} \qquad (8.19)$$

where

$$E_N(T') = \frac{(\pi k T')^2}{6h} \qquad (8.20)$$

The total output radiation power is $P = S + N$, which can be expressed as

$$\frac{(\pi k T_e)^2}{6h} = S + \frac{(\pi k T)^2}{6h} \qquad (8.21)$$

Therefore one can evaluate T_e as a function of signal power S and noise temperature T:

$$T_e = \left[\frac{6hS}{(\pi k)^2} + T^2 \right]^{1/2} \qquad (8.22)$$

By substituting Eq. (8.22) into Eq. (8.19), the capacity of a photon channel can be shown:

$$C = \frac{\pi^2 k T}{3h \ln 2} \left[\left(1 + \frac{6hS}{(\pi k T)^2} \right)^{1/2} - 1 \right] \qquad (8.23)$$

Since the amount of energy required to transmit 1 bit of information is S/R, where R is the rate of information, we have

$$\frac{S}{R} = \frac{kTC \ln 2}{R} + \frac{3hC^2(\ln 2)^2}{2\pi^2 R} \qquad (8.24)$$

It is also noted that the rate of information can never be greater than the channel capacity $(R \leq C)$, thus the minimum signal energy required to transmit the information at a rate equal to the channel capacity $(R = C)$ is

$$S_{\min} = kT \ln 2 + \frac{3}{2\pi^2} hR(\ln 2)^2 \qquad (8.25)$$

Although the signal spectrum is excluded to infinity, it decreases rapidly at $h\nu \gg kTe$ [according to Eq. (8.17)]. Thus the bandwidth required to obtain the channel capacity of Eq. (8.23) can be estimated:

$$\Delta \nu \simeq \frac{kT_e}{h} = \frac{kT}{h}\left[\frac{6hS}{(\pi kT)^2} + 1\right]^{1/2} \tag{8.26}$$

It is noted that, if the signal-to-noise ratio is high ($h\nu \gg kT$):

$$\frac{6hS}{(\pi kT)^2} \gg 1 \tag{8.27}$$

then the channel capacity of Eq. (8.23) is limited by the quantum statistic,

$$C_{\text{quant}} = \frac{\pi}{\ln 2}\left(\frac{2S}{3h}\right)^{1/2} \tag{8.28}$$

We see that this quantum channel capacity agrees with Stern's [10] result, except by a factor of $\sqrt{2}$.

However if the signal-to-noise ratio is low ($h\nu \ll kT$):

$$\frac{6hS}{(\pi kT)^2} \ll 1 \tag{8.29}$$

then the channel capacity of Eq. (8.23) is reduced to the classical limit:

$$C_{\text{class}} = \frac{S}{kT\ln 2} \tag{8.30}$$

Equation (8.30) is essentially Shannon's [1, 3] result for a wide-band channel with additive Gaussian noise.

It is also noted that, if the rate of information is low, then the minimum signal energy required [Eq. (8.25)] can be approximated by the equation

$$S_{\min} \simeq kT\ln 2 \tag{8.31}$$

This is the result we obtained in a previous chapter [Eq. (6.16)] for low-frequency observation.

It is also interesting to plot the capacity of the photon channel of Eq. (8.23) as a function of S for various values of thermal noise temperature T, as shown in Fig. 8.1. We see that the capacity of the channel depends on the signal-to-noise ratio. The classical limit approaches the quantum limit asymptotically, at a point of intersection, where

$$\frac{6hS}{(\pi kT)^2} = 4 \tag{8.32}$$

The photon channel for a low signal-to-noise ratio agrees with the

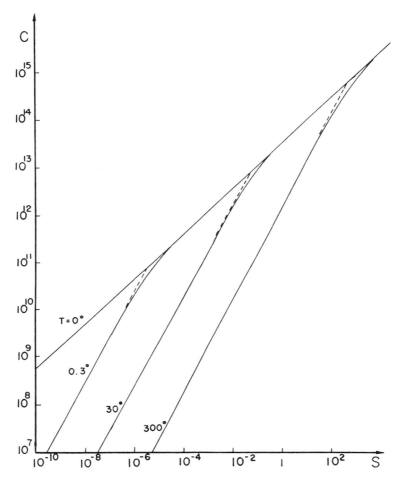

Fig. 8.1 Capacity of channel for radiation as a function of signal power, for various values of thermal noise temperature T. Dashed lines represent the classical asymptotes of Eq. (8.30).

classical limit. However, for a high signal-to-noise ratio, the quantum channel of Eq. (8.28) appears much higher in comparison to the typical information rate in practice. This is essentially due to the idealized transmitter and receiver we have assumed in the calculation. In the following sections, a more detailed discussion of the photon channel is given, and in particular is concentrated on narrow-band channels.

8.3 AN INFORMATIONAL THEORISTIC APPROACH

In Sec. 8.2 we calculated the capacity of a photon channel from the thermodynamic entropy standpoint. An identical result can also be derived from a strictly informational theoristic point of view. However, we will see that the latter approach is more mathematically inclined.

Let us now suppose that, at a finite time interval of Δt, a signal embedded in additive noise is received. It is assumed that the received signal is located within a small localized spherical volume with a radius proportional to Δt. The photons of the received signal are assumed to have a discrete spectral energy:

$$E_i = h\nu_i \qquad i = 1, 2, \ldots \tag{8.33}$$

where $\nu_i = i/\Delta t$.

Let the input signal mean occupational quantum number be m_i, at frequency ν_i, and let the mean occupational quantum number of the noise be n_i. Thus the output signal (signal plus noise) for every ν_i can be described as

$$f_i = m_i + n_i \tag{8.34}$$

Let us denote the corresponding probability distributions for m_i, n_i, and f_i by $p(m_i)$, $p(n_i)$, and $p(f_i)$, respectively. Since m_i and n_i are assumed to be statistically independent, the corresponding conditional probability can be written

$$p(f_i/m_i) = p(f_i - m_i) \tag{8.35}$$

The average signal energy density per unit time is therefore

$$E_S = \frac{1}{\Delta t} \sum_{i=0}^{\infty} \sum_{m_i=0}^{\infty} m_i P(m_i) h\nu_i \tag{8.36}$$

Thus the amount of mutual information between the input and the output radiation (photon channel) for every ν_i is

$$I(f_i; m_i) = H(f_i) - H(f_i/m_i) \tag{8.37}$$

Since the signal and the noise are assumed to be statistically independent, it is clear that

$$H(f_i/m_i) = f(n_i) \tag{8.38}$$

Thus the corresponding mutual information of the photon channel is

$$I(f_i; m_i) = H(f_i) - H(n_i) \tag{8.39}$$

By the substitution of the definition of $H(f_i)$ and $H(n_i)$ in Eq. (8.39), Eq. (8.39) can be written

$$I(f_i; m_i) = \sum_{m_i=0}^{\infty} \sum_{f_i=m_i}^{\infty} p(m_i)p(f_i/m_i) \log_2 \frac{p(f_i/m_i)}{p(f_i)} \tag{8.40}$$

Thus the total amount of mutual information per unit time is the ensemble average of $I(f_i; m_i)$ over i:

$$I(f; m) = \frac{1}{\Delta t} \sum_{i=0}^{\infty} I(f_i; m_i) \tag{8.41}$$

and it can be shown that

$$I(f; m) = \frac{1}{\Delta t} [H(f) - H(n)] \tag{8.42}$$

Let us recall that the photon fluctuation (due to thermal radiation) follows Gibbs' distribution:

$$p(n_i) = \frac{1 - \exp(-h\nu_i/kT)}{\exp(n_i h\nu_i/kT)} \tag{8.43}$$

Then the corresponding conditional entropy equivocation can be written

$$\begin{aligned} H(n_i) &= \sum_{n_i=0}^{\infty} p(n_i) \log_2 p(n_i) \\ &= \frac{h\nu_i}{kT[\exp(h\nu_i/kT) - 1]\ln 2} - \log_2[1 - \exp(-h\nu_i/kT)] \end{aligned} \tag{8.44}$$

Since $I(f; m)$ is maximum whenever $H(f)$ is maximum, we seek the distribution of $p(f_i)$ for which $H(f)$ is maximum. It is noted that the average received signal power can be written

$$\frac{1}{\Delta t} \sum_{i=1}^{\infty} \sum_{f_i=1}^{\infty} f_i h\nu_i p(f_i) = S + \frac{1}{\Delta t} \sum_{i=1}^{\infty} \frac{h\nu_i}{\exp(h\nu_i/kT) - 1} \tag{8.45}$$

where S is the corresponding signal power and $p(f_i)$ is the probability distribution of the received signal:

$$\sum_{f_i=1}^{\infty} p(f_i) = 1 \tag{8.46}$$

Thus the maximum entropy can be obtained by Gibbs' distribution:

$$p(f_i) = \frac{1 - \exp(-h\nu_i/\alpha)}{\exp(f_i h\nu_i/\alpha)} \tag{8.47}$$

where α is a Lagrange multiplier having a thermal energy of $\alpha = kT_e$. We now let the time interval approach infinity, $\Delta t \to \infty$, and then $1/\Delta t$ can be written as an incremental frequency element, $1/\Delta t = dv$. Thus the noise entropy of Eq. (8.44) can be obtained by the integral

$$H(n) = \int_0^\infty \left\{ \frac{hv}{[\exp(hv/kT) - 1]kT \ln_2} - \log_2[1 - \exp(-hv/kT)] \right\} dv \tag{8.48}$$

which can be reduced to

$$H(n) = \frac{\pi^2 kT}{3h \ln_2} \tag{8.49}$$

Similarly, the output signal entropy can be determined:

$$H(f) = -\sum_{f_i} \int_0^\infty p(f_i) \log_2 p(f_i) \, dv \tag{8.50}$$

By substituting Eq. (8.47) into Eq. (8.50), one can show that

$$H(f) = \frac{\pi^2 \alpha}{3h \ln 2} \tag{8.51}$$

It is noted that α (the Lagrange multiplier) equals kT_e, as in the previous section. By imposing the condition of Eq. (8.45), one could have obtained the photon channel capacity of Eq. (8.23):

$$C = H(f) - H(n) = \frac{\pi^2 kT}{3h \ln 2} \left[\left(1 + \frac{6hS}{(\pi RT)^2}\right)^{1/2} - 1 \right] \tag{8.52}$$

which is identical to Eq. (8.23).

It is interesting to obtain the explicit form of the probabilistic signal distribution. Since it is assumed that the signal and the noise are statistically independent, we have the following equation for every v_i:

$$p(f_i) = \sum_{m_i=0}^{f_i} p(m_i) p(f_i - m) \tag{8.53}$$

where $n = f - m$, and $p(m)$ and $p(n)$ denote the signal and noise probability distribution, respectively. (*Note*: To simplify the notation, the subscript of i is omitted in the following equations.)

If we let the Lagrange multiplier $\alpha = kT_e$, and substitute Eqs. (8.43) and (8.47) into Eq. (8.53), we then obtain the equations

$$p(m)|_{m=0} = \frac{1 - \exp(-h\nu/kT_e)}{1 - \exp(-h\nu/kT)} \tag{8.54}$$

and

$$p(m) = \frac{[1 - \exp(-h\nu/kT_e)]\{1 - \exp[(h\nu/k)(1/T_e - 1/T)]\}}{[1 - \exp(-h\nu/kT)]\exp(mh\nu/kT_e)} \tag{8.55}$$

We see that the signal distribution resembles, but is quite different from, Gibbs' distribution. Thus it is incorrect to assume that the signal is statistically similar to that of thermal noise $P(n)$.

The mean occupational quantum number (the signal) can therefore be obtained by the definition

$$\bar{m} = \sum_{m=1}^{\infty} m p(m) = \frac{1}{\exp(h\nu/kT_e) - 1} - \frac{1}{\exp(h\nu/kT) - 1} \tag{8.56}$$

Thus we see that the signal as a function of ν does not follow the Bose-Einstein distribution, in contrast to the normal assumption [12].

8.4 NARROW-BAND PHOTON CHANNEL

Let us now propose a narrow-band photon channel in which the effective bandwidth $\Delta\nu$ is assumed to be very narrow as compared with its center frequency ν, $(\Delta\nu \ll \nu)$. In other words, the noise power is mostly concentrated within a narrow bandwidth of $\Delta\nu$. Thus the corresponding power spectral density can be considered approximately uniform throughout the narrow bandwidth $\Delta\nu$ and can be approximated:

$$N(\nu, T) \simeq \frac{h\nu}{\exp(h\nu/kT) - 1} \tag{8.57}$$

The narrow-band noise power as a function of temperature T can be written

$$N(T) = N(\nu, T)\Delta\nu \tag{8.58}$$

It is noted that, as in the previous sections, the maximum information rate can be achieved if the output signal (signal plus noise) is also a Gibbs' distribution of Eq. (8.47). Thus Eq. (8.58) provides an optimum transfer of information for a narrow-band channel. Since it is assumed that photon noise is additive, the output power spectral density is

$$P(v, T) = \frac{S}{\Delta v} + N(v, T) \tag{8.59}$$

where $S/\Delta v$ is the signal power spectral density.

Again, based on the entropy theory of information, as shown in Eq. (8.19), the capacity of a narrow-band photon channel can be determined by the integral

$$C = \frac{\Delta v}{k \ln 2} \int_N^P \frac{d[N(v, T)]}{T(N, v)} \tag{8.60}$$

By evaluating the above equation, one can show that

$$C = \Delta v \left\{ \log_2 \left[1 + \frac{S}{hv \Delta v}(1 - e^{-hv/kT}) \right] + \left(\frac{S}{hv \Delta v} + \frac{1}{e^{hv/kT} - 1} \right) \right.$$
$$\left. \cdot \log_2 \left[1 + \frac{hv \Delta v(e^{hv/kT} - 1)}{S(e^{hv/kT} - 1) + hv \Delta v} \right] - \frac{hv/kT}{\ln 2(e^{hv/kT} - 1)} \right\} \tag{8.61}$$

Equation (8.61) is essentially the result obtained by Gordon [12, eq. (5)].

From Eq. (8.61), we see that the channel capacity decreases monotonically with the frequency v. The monotonic behavior $C/\Delta v$, as a function of hv/kT, is plotted in Fig. 8.2 for various values of $S/kT \Delta v$.

Furthermore, if one lets the mean occupational quantum number $\bar{m}$ remain fixed, then the minimum number of quanta required for 1 bit of information to be transfered can be written

$$M = \frac{\bar{m} \Delta v}{C} \tag{8.62}$$

From Eq. (8.62), we see that the number of quanta remains uniform at $hv \gg kT$ (the quantum effect) but increases rapidly as the frequency decreases, $hv \ll kT$. The effect of M, the minimum required quanta, as a function of hv/kT is plotted in Fig. 8.3, for various values of $\bar{m}$. It is also interesting to note that, if $hv \ll kT$, then the narrow-band photon channel capacity can be reduced to the approximation

$$C \simeq \Delta v \log_2 \left(1 + \frac{S}{N} \right) \quad \text{for } hv \ll kT \tag{8.63}$$

where $N = kT \Delta v$ is the thermal noise power of the channel. Equation (8.63) essentially represents the well-known Gaussian channel described by Shannon [1]. Let us recall that, when Shannon derived this expression for a Gaussian channel, he completely disregarded the physical nature of the problem. Nevertheless, his Gaussian channel is the asymptotic result

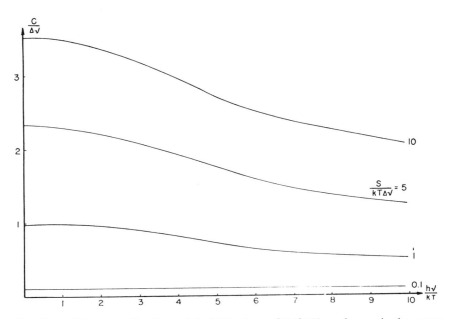

Fig. 8.2 $C/\Delta v$ as a function of hv/kT values of $S/kT\Delta v$, where v is the center frequency and Δv is the bandwidth of the narrow-band channel.

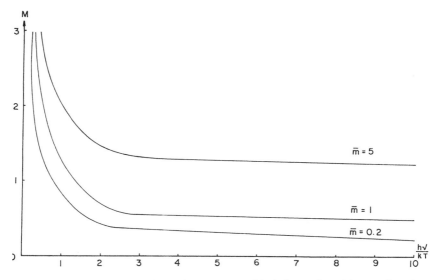

Fig. 8.3 Minimum member of quanta per bit information M as a function of hv/kT, for various values of m.

of our quantum channel of Eq. (8.61) at the low-frequency extreme ($hv \ll kT$). In other words, at low frequency ($hv \ll kT$), the quantum energy of hv can be approximated by a continuous energy spectrum for which the amplitude of the radiating field can be approximated by a Gaussian statistic. It is well known that Gaussian amplitude distribution corresponds to continuous exponential energy distribution, which follows Gibbs' statistics. Moreover we see that, as the limit of $hv/kT \to 0$, the Gibbs' distribution for the noise [Eq. (8.43)] becomes

$$p(E_n) = \frac{1}{kT} \exp\left(-\frac{E_n}{kT}\right) \tag{8.64}$$

where $E_n = nhv$. Similarly, from Eq. (8.47), as $hv/kT \to 0$, the output signal energy distribution (signal plus noise) can be shown:

$$p(E_f) = \frac{1}{kT_e} \exp\left(-\frac{E_f}{kT_e}\right) \tag{8.65}$$

where $E_f = fhv$, $kT_e = S + N$ (the total output signal power), and T_e is the *effective output signal temperature*.

As noted in Chap. 2, a time-variable signal (in our case a one-dimensional polarized electromagnetic signal or photon beam) with bandwidth Δv and duration. Δt, where $\Delta v \, \Delta t \gg 1$, can be completely described by $2 \Delta v \Delta t$ degrees of freedom. However, from the quantum theory point of view, the electromagnetic field is primarily localized in a *phase volume* of $h \Delta v \, \Delta t$, and the signal can be described *only* by $\Delta v \, \Delta t$ degrees of freedom, namely, the energy values assigned to various quantum levels. It is clear that the minimum sampling frequency, from the quantum standpoint, requires only $1/\Delta t$, instead of $1/2 \, \Delta t$. Let us now assume that the noise amplitudes in the x and y (vertical and horizontal) directions are statistically independent and can be described by zero-mean Gaussian statistics, with the respective variances $\sigma_x^2 = \sigma_y^2 = N/2$:

$$p(x) = \frac{1}{\sqrt{\pi N}} \exp\left(-\frac{x^2}{N}\right) \tag{8.66}$$

and

$$p(y) = \frac{1}{\sqrt{\pi N}} \exp\left(-\frac{y^2}{N}\right) \tag{8.67}$$

where N is the average noise power. It can be seen that the corresponding noise energy distribution, $E_n = x^2 + y^2$, follows Gibbs' distribution:

$$p(E_n) = \frac{1}{kT} \exp\left(-\frac{E_n}{kT}\right) \tag{8.68}$$

which results from Eq. (8.64).

Similarly, we let the output signal (signal plus noise) amplitudes in the x and y directions have Gaussian distributions, with their variances equal to $(S+N)/2$. Thus we have

$$p_f(x) = \frac{1}{\sqrt{\pi(S+N)}} \exp\left(-\frac{x^2}{S+N}\right) \tag{8.69}$$

and

$$p_f(y) = \frac{1}{\sqrt{\pi(S+N)}} \exp\left(-\frac{y^2}{S+N}\right) \tag{8.70}$$

where S is the input signal power. We see again that the output signal energy, $E_f = x^2 + y^2$, follows Gibbs' statistics:

$$p(E_f) = \frac{1}{kT_e} \exp\left(-\frac{E_f}{kT_e}\right) \tag{8.71}$$

where is identical to Eq. (8.65).

Now we calculate the amount of mutual information (information transfer) per degree of freedom, first from the classical theory point of view:

$$\begin{aligned} I_{\text{class}} &= H(f) - H(n) \\ &= -\int_{-\infty}^{\infty} p_f(x) \log_2 p_f(x)\, dx + \int_{-\infty}^{\infty} p_n(x) \log_2 p_n(x)\, dx \end{aligned} \tag{8.72}$$

By substituting Eqs. (8.69) and (8.68) into Eq. (8.72) we obtain

$$I_{\text{class}} = \tfrac{1}{2} \log_2\left(1 + \frac{S}{N}\right) \tag{8.73}$$

which was the formula obtained by Shannon [1].

Similarly, from the quantum theory standpoint, the amount of information transfer per degree of freedom can be determined:

$$\begin{aligned} I_{\text{quant}} &= -\int_0^{\infty} p(E_f) \log_2 p(E_f)\, dE_f + \int_0^{\infty} p(E_n) \log_2 p(E_n)\, dE_n \\ &= \log_2\left(1 + \frac{S}{N}\right) \end{aligned} \tag{8.74}$$

Therefore we see that

$$I_{\text{quant}} = 2 I_{\text{class}} \tag{8.75}$$

The amount of information at the quantum limit is twice as large as that at the classical limit. But the degrees of freedom in the classical theory is 2 $\Delta v \, \Delta t$, which is twice that for the quantum limit. Therefore we conclude that the capacity of the quantum-mechanical channel is equal to that of the classical channel:

$$C = 2 \Delta v \, I_{\text{class}} = \Delta v \, I_{\text{quant}} = \Delta v \log_2 \left(1 + \frac{S}{N} \right) \tag{8.76}$$

This interesting result may explain certain ambiguities and the paradoxes of the classical channel of Eq. (8.1) that Shannon calculated. We have shown that the quantum-mechanical channel coincides with the classical channel asymptotically at the lower frequency limit, as $hv/kT \to 0$. It is noted that the quantum effect takes place in the communication channel when

$$\frac{hv}{kT} \gg 1 \tag{8.77}$$

and

$$\frac{S}{hv \, \Delta v} \left[\exp\left(\frac{hv}{kT}\right) - 1 \right] \gg 1 \tag{8.78}$$

that is, when the mean occupation quantum number of the thermal noise is not only much smaller than unity but is also very small as compared with the mean occupation quantum number of the input signal. Thus we see that, under the conditions of Eqs. (8.77) and (8.78), Eq. (8.61) can be approximated:

$$C \simeq \Delta v \left[\log_2 \left(1 + \frac{S}{hv} \right) + \frac{S}{hv} \log_2 \left(1 + \frac{hv}{S} \right) \right] \tag{8.79}$$

The capacity of the channel increases as the mean occupation number $\bar{m} = S/hv$ increases. But it should be cautioned that Eq. (8.79) is valid only under the condition $\Delta v/v \ll 1$, that is, for a narrow-band channel. Therefore it is incorrect to assume that the capacity of the channel becomes infinitely large when $v \to 0$; strictly speaking, the capacity will never exceed the quantum limit as given in Eq. (8.28).

Furthermore, for high signal power S, which implies that $S/h\nu\,\Delta\nu \gg 1$, Eq. (8.61) can be approximated by Shannon's formula:

$$C \simeq \Delta\nu \log_2\left(1 + \frac{S}{kT_{eq}}\right) \tag{8.80}$$

where $kT_{eq} = h\nu/e, e = 2.718\ldots$, and T_{eq} is the *equivalent noise temperature*.

It must be pointed out that, according to Einstein's formula for energy fluctuations in blackbody radiation, the equivalent temperature of the quantum noise is defined as $kT_{eq} = h\nu$. But according to the generalized Nyquist formula, the definition of T_{eq} is $kT_{eq} = h\nu/2$. For the problem quantum effect in information transmission, according to Lebedev and Levitin [13], T_{eq} is defined as $kT_{eq} = h\nu/e$.

In concluding this section we point out that, in the case of two possible independent polarized states of radiation, the photon channel we have discussed can be considered two independent channel. If the total input signal power is assumed constant, then the overall capacity (i.e., the combination of these two channels) will be maximum if and only if the input signal power is equally divided between the channels. To obtain the overall channel capacity, one can simply replace the signal power S by $S/2$ in all the capacity equations in the previous sections and multiply the whole equation by a factor of 2. We see that in the quantum limit case of Eq. (8.28) the use of the two possible polarizations gives rise to a 2-fold gain in capacity. However, in the classical limit case of Eq. (8.30), we see that the channel capacity does not change after the modification.

8.5 OPTIMUM SIGNAL POWER DISTRIBUTION, A SPECIAL CASE

We illustrate the problem of distributing the signal power over the bandwidth of the channel to obtain the optimum capacity of a photon channel. For example, take the case of noise temperature as a function of frequency, $T = T(\nu)$.

To approach this distribution problem, we adopt the expression for the narrow-band channel given in Eq. (8.61). We see that this overall capacity can be obtained by the expression

$$C_0 = \int_0^\infty \frac{C(\nu)}{\Delta\nu} d\nu \tag{8.81}$$

where $C(\nu)$ is the capacity of the narrow-band channel, which is a function of the center frequency ν. It is also noted that the signal power density and the noise temperature are also functions of the frequency ν [$S(\nu)$ and $T(\nu)$].

Quantum Effect on Information Transmission 213

Since we assume that the photon noise is additive, the output power spectral density of the channel is the sum of the signal and noise spectral densities:

$$P(v) = S(v) + N(v) \tag{8.82}$$

where $S(v)$ and $N(v)$ are the signal and noise spectral densities, respectively. The noise power spectral density can be described by the equation

$$N(v) = \frac{hv}{\exp[hv/kT(v)] - 1} \tag{8.83}$$

In order to obtain an optimum power distribution of the output signal (signal plus noise), one can apply the variational technique to the integral of Eq. (8.81) under the following conditions:

$$\int_0^\infty P(v)\,d(v) = S + \int_0^\infty \frac{hv\,dv}{\exp[hv/kT(v)] - 1} \tag{8.84}$$

where S is the input signal power.

Since the output power spectral density $P(v)$ is at least equal to or higher than that of the input,

$$P(v) \geq S(v) \tag{8.85}$$

at the locations where noise spectral density is higher than that of $S(v)$, $P(v)$ follows the variation in the noise spectral density $N(v)$. Thus the distribution of the optimum output power spectral density (signal plus noise) is

$$P(v) = \begin{cases} \dfrac{hv}{\exp(hv/kT_e) - 1} & T_e \geq T(v) \\ N(v) & T_e < T(v) \end{cases} \tag{8.86}$$

where T_e is the equivalent temperature. It is clear that Eq. (8.86) is essentially the result we obtained previously. The effective output signal temperature T_e can be determined by Eq. (8.84):

$$S = \int_I [P(v) - N(v)]\,dv \tag{8.87}$$

where I indicates integration over the interval for which $T_e \geq T(v)$.

Thus we see that the optimum spectral distribution of the input signal power is such that all the degrees of freedom of the received signal are *heated* (or raised) to the same noise temperature T, as shown in Fig. 8.4.

Moreover, in classical theory ($hv \ll kT$) the output power spectral

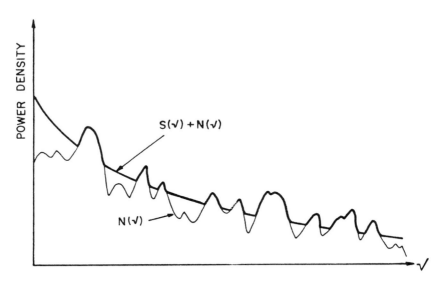

Fig. 8.4 Optimum distribution of signal power spectral, a special case of frequency depending noise temperature $T(v)$. The lower curve represents the noise power spectral and the upper curve represents the output signal (i.e., signal plus noise) power spectral.

density is simply proportional to the temperature T, that is, the output power remains constant:

$$S + N = \text{constant} \tag{8.88}$$

We see that Eq. (8.88) is essentially the result Shannon obtained.

In this chapter we have treated in detail the basic quantum effect on information transmission. For such a quantum channel, for maximum entropy transfer, the output signal ensemble has been shown to follow Gibbs' distribution, for which it corresponds to the thermal equilibrium. Thus by means of the entropy theory of information, the capacity of a photon channel can be calculated. As a result, we have shown that, although the information transfer at the quantum limit is twice as large as that at the classical limit, the corresponding degree of freedom at the quantum limit is a half that at the classical. Therefore the capacity of the quantum channel should remain equal to that of the classical channel. We have also shown that, as the frequency is lowered ($hv \ll kT$), this information capacity of the photon channel approaches asymptotically the continuous additive Gaussian channel of Shannon. As the frequency approaches infinity, the photon channel capacity approaches a finite quantum limit. Furthermore, for a frequency-dependent noise temperature, it is possible at least in theory to obtain an optimum channel by distributing the signal power spectral over the frequency domain.

REFERENCES

1. C. E. Shannon and W. Weaver, *The Mathematical Theory of Communication*, University of Illinois Press, Urbana, 1949.
2. L. Szilard, "Über die Entropieverminderung in Einem Thermodynamischen System bei Eingriffen Intelligenter Wessen," *Z. Phys.*, vol. 53, 840 (1929).
3. C. E. Shannon, "A Mathematical Theory of Communication," *Bell. Syst. Tech. J.*, vol. 27, 379–423, 623–656 (1948).
4. D. Gabor, "Communication Theory and Physics," *Phil. Mag*, vol. 41, no. 7, 1161 (1950).
5. L. Brillouin, "Thermodynamics and Information Theory," *Am Sci.*, vol. 38, 594 (1950).
6. L. Brillouin, "Maxwell's Demon Cannot Operate. Information Theory and Entropy I," *J. Appl. Phys.*, vol. 22, 334 (1951).
7. L. Brillouin, "Physical Entropy and Information II," *J. Appl. Phys.* vol. 22, 338 (1951).
8. L. Brillouin, "The Negentropy Principle of Information," *J. Appl. Phys.*, vol. 24, 1152 (1953).
9. L. Brillouin, *Science and Information Theory, Academic*, New York, 1956.
10. T. E. Stern, "Some Quantum Effects in Information Channels," *IRE Trans.*, IT-6, no. 4, 435 (1960).
11. T. E. Stern, "Information Rates in Photon Channels and Photon Amplifiers," *IRE Int. Conv. Rec.*, pt. 4, 182 (1960).
12. J. P. Gordon, "Quantum Effects in Communication Systems," *Proc. IRE*, vol. 50, 1898 (1962).
13. D. S. Lebedev and L. B. Levitin, "Information Transmission by Electromagnetic Field," *Inf. Control*, vol. 9, 1 (1966).
14. D. Gabor, "Light and Information," in E. Wolf, ed., *Progress in Optics*, vol. I, North-Holland, Amsterdam, 1961.

9
Coherence Theory of Optics

9.1 ASPECTS OF COHERENCE

In the widespread application of coherent light, it is appropriate to discuss the principles of coherence in radiation. If the radiations from two point sources maintain a fixed phase relation between them, they are said to be *mutually coherent*. An extended source is coherent if all points of the source have fixed phase differences between them.

In classical electromagnetic radiation, it is usually assumed that the electric and magnetic fields are always measurable at any time and position. In these cases, no accounting of coherence theory is needed. However, there are situations that this assumption cannot be made and it is often helpful to apply coherence theory in the problem. For example, if we want to determine the diffraction pattern caused by the radiation from several sources, we cannot obtain an exact result unless the degree of coherence among the sources are known. In such a situation, however, it is desirable to obtain an ensemble average to represent the *statistically* most likely result from any such combination of sources. It may thus be more useful to provide a statistical description than to follow the dynamical behavior of a wave field in detail.

Let us assume an electromagnetic wave field propagating in space, as depicted in Fig. 9.1, where $u_1(t)$ and $u_2(t)$ denote the instantaneous wave disturbances at positions 1 and 2, respectively. We choose the second-order moment as the quantity to be averaged, as given by

$$\Gamma_{12}(\tau) \triangleq \langle u_1(t+\tau)u_2^*(t)\rangle \tag{9.1}$$

which is known as the *mutual coherence function*, where the asterisk denotes the complex conjugate and τ denotes a time delay variable. The angle brackets represents the time ensemble as written by

$$\Gamma_{12}(\tau) = \lim_{T\to\infty} \frac{1}{2T} \int_{-T}^{T} u_1(t+\tau)u^*(t)\,dt \tag{9.2}$$

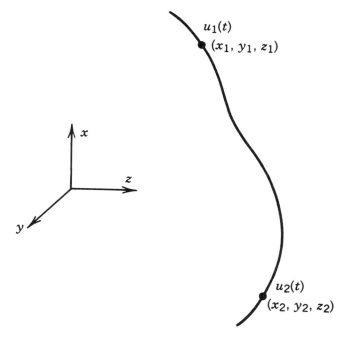

Fig. 9.1 A complex wavefront in space.

It is apparent that the mutual coherence function is essentially the temporal *cross-correlation function* between $u_i(t)$ and $u_2(t)$.

The complex degree of coherence can be defined as

$$\gamma_{12}(\tau) \triangleq \frac{\Gamma_{12}(\tau)}{[\Gamma_{11}(0)\Gamma_{22}(0)]^{1/2}} \tag{9.3}$$

in which $\Gamma_{11}(\tau)$ and $\Gamma_{22}(\tau)$ are the *self-coherence functions* of $u_1(t)$ and $u_2(t)$, respectively.

The normalized mutual coherence function $\gamma_{12}(\tau)$ can be measured by applying Young's experiment, as shown in Fig. 9.2, in which Σ represents an extended monochromatic source. As light falls upon a diffraction screen at a distance r_{10} from the source, with two small pinholes Q_1 and Q_2 separated by a distance d at the screen, an interference pattern is observed at the observation screen. We shall adopt a measurable quantity, defined by *Michelson's visibility* of the fringes, as given by

$$v \triangleq \frac{I_{\max} - I_{\min}}{I_{\max} + I_{\min}} \tag{9.4}$$

Coherence Theory of Optics

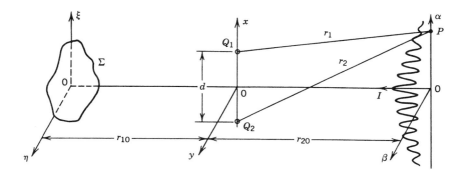

Fig. 9.2 Young's experiment. The Σ is an extended but nearly monochromatic source.

where I_{max} and I_{min} are the maximum and the minimum intensities of the fringes.

As the source size Σ shrinks, we see that the visibility of the fringes increases. By varying the distance d between the pinholes, the visibility varies as d increases, as shown in Fig. 9.3. The variation of visibility can be in fact used as a measure of *spatial coherence* on the diffraction screen.

On the other hand, as the point of observation P moves away from the center of the observing screen, visibility decreases as the path difference $\Delta r = r_1 - r_2$ increases, until it becomes zero. This effect also depends on the monochromaticity of the source. The visibility of the fringes has been found to be appreciable only within a path difference of

$$\Delta r \approx \frac{2\pi c}{\Delta \omega} \tag{9.5}$$

which is known as the *coherent length* of the source, where c denotes the light velocity and $\Delta \omega$ is the spectral width of the source.

It is, however, not necessary to have completely coherent light to produce an interference pattern, but under the right conditions such as a pattern may be obtained from an incoherent source. This effect is called *partial coherence*.

Since $u_1(t)$ and $u_2(t)$ can be subjected to the scalar wave equation as given by

$$\nabla^2 u = \frac{1}{c^2} \frac{\partial^2 u}{\partial u^2} \tag{9.6}$$

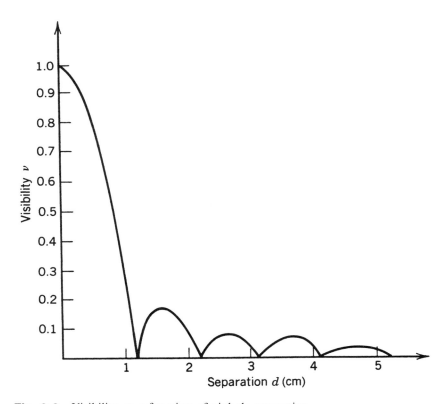

Fig. 9.3 Visibility as a function of pinhole separation.

the intensity at P can be shown as

$$I_P = \langle u_P(t)u_P^*(t)\rangle = I_1 + I_2 + 2\,\mathrm{Re}\left\langle c_1 u_1\left(t - \frac{r_1}{c}\right) c_2^* u_2^*\left(t - \frac{r_2}{c}\right)\right\rangle \tag{9.7}$$

where I_1 and I_2 are the intensities from $u_1(t)$ and $u_2(t)$ and $u_2(t)$, respectively.

By letting $t_1 = r_1/c$ and $t_2 = r_2/c$, and $\tau = t_1 - t_2$, the preceding equation becomes

$$I_p = I_1 + I_2 + 2c_1 c_2^*\,\mathrm{Re}\,[\Gamma_{12}(\tau)] \tag{9.8}$$

By allowing $|c_1|^2 \Gamma_{22}(0) = I_2$, and $|c_1|^2 \Gamma_{11}(0) = I_1$, called at the *best condition*, the intensity can be written as

$$I_p = 2I[1 + |\gamma_{12}(\tau)|\cos\phi_{12}(\tau)] \tag{9.9}$$

Coherence Theory of Optics

in which we see that

$$v = |\gamma_{12}(\tau)| \tag{9.10}$$

In other words, under the best conditions the visibility of the fringes is a measure of the degree of coherence.

9.2 SPATIAL AND TEMPORAL COHERENCE

In the preceding section, we saw that the *term spatial coherence* is applied to the effects due to the source size. For example, if we look at two points that are at equal light path distances from a point source, the radiations reaching these points will be exactly the same. Then the mutual coherence will be equal to the self-coherence at either point, that is,

$$\Gamma_{12}(Q_1, Q_2, \tau) = \langle u(Q_1, t+\tau) u^*(Q_2, t) \rangle = \Gamma_{11}(\tau) \tag{9.11}$$

As the source is enlarged, we can no longer claim an equality of mutual coherence and self-coherence. This lack of complete coherence is a *spatial* effect. On the other hand, the *temporal coherence* is an effect that is due to the finite spectral width of the source. The coherence is complete for strictly monochromatic radiation but becomes only partial as other wavelengths are added. It is never possible to separate completely spatial and temporal coherence, but it is well to name them and point out their significances.

9.2.1 Spatial Coherence

By utilizing Young's experiment, the angular size of a spatially incoherent source (e.g., an extended source) as it relates to spatial coherence can be determined. Let us now consider the paths of two light rays from a point on a linearly extended source ΔS, as they pass through the two narrow slits Q_1 and Q_2 depicted in Fig. 9.4. If we let $r_{10} \gg d$, the intensity distribution at observation screen P can be written as

$$I_p(\alpha) = I_1 + I_2 + 2\sqrt{I_1 I_2} \cos\left[k\left(\frac{d}{r_{10}}\xi + \frac{d}{r_{20}}\alpha\right)\right] \tag{9.12}$$

which describes a set of parallel fringes along the α axis. For a

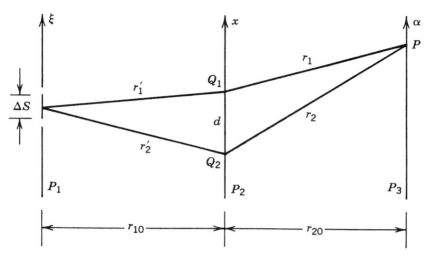

Fig. 9.4 Measurement of spatial coherence.

one-dimensional extended source of size ΔS, the resulting fringe pattern can be shown as

$$I'_p(\alpha) = I_1 \Delta S + I_2 \Delta S + 2\sqrt{I_1 I_2} \int_{\Delta S} \cos\left[k\left(\frac{d}{r_{10}}\xi + \frac{d}{r_{20}}\alpha\right)\right] d\xi$$

$$= \Delta S \left\{ I_1 + I_2 + 2\sqrt{I_1 I_2} \operatorname{Sinc}\left[\left(\frac{\pi d}{\lambda r_{10}}\right)\Delta S\right] \cos\left(k\frac{d}{r_{20}}\alpha\right) \right\}$$

(9.13)

as the fringes vanish at

$$d = \frac{\lambda r_{10}}{\Delta S} \qquad (9.14)$$

Thus, we see that region of spatial coherence is determined by the visibility of the interference fringes. We see that the coherence region increases as the source size decreases or as the distance of the source increases. In other words, the degree of spatial coherence increases as the distance of the propagated wave increases.

9.2.2 Temporal Coherence

It is possible to split a light wave into two paths, to delay one of them, and then to recombine them to form an interference fringe pattern. In this way, we can measure the temporal coherence of the light wave. In other words, the degree of temporal coherence is the measure of the cross-correlation

Coherence Theory of Optics

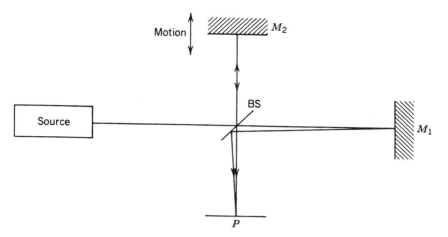

Fig. 9.5 Measurement of temporal coherence. BS, beam splitter; M, mirrors, P, observation screen.

of a wave field at one time with respect to another wave field at a later time. Thus, the definition of temporal coherence can also refer to *longitudinal coherence* as opposed to *transversal* (i.e., spatial) coherence. The maximum difference in the optical path lengths of two waves derived from a source is known as the *coherent length* of that source. Since spatial coherence is determined by the wave field in the transversal direction and temporal coherence is measured along the longitudinal direction, spatial coherence and temporal coherence described the degree of coherence of a wave field within a spatial volume.

The *Michelson interferometer* is one of the techniques commonly used to measure the temporal coherence of a light source, as shown in Fig. 9.5. The beam splitter BS divides the light beam into two paths. one of which goes to mirror M_1 and the other goes to mirror M_2. By varying the path length (i.e., τ) to one of the mirrors, the variation of the interference fringes at screen P can be observed. Thus, the measurement of the visibility of the interference fringes corresponding to the measurement of the degree of temporal coherence $\gamma_{11}(\tau)$, at a time delay τ, can be obtained.

Since the physical nature of a light source affects the coherency of wave propagation, the spectral width of the source affects the temporal coherence. The time interval Δt, during which the wave propagated is coherent, can be approximated by

$$\Delta t \approx \frac{2\pi}{\Delta \omega} \tag{9.15}$$

for which the coherence length of the light source can be written as

$$\Delta r = \Delta t c \approx \frac{2\pi c}{\Delta \omega} \tag{9.16}$$

where c is the velocity of light, and $\Delta \omega$ is the spectral bandwidth of the light source.

9.3 COHERENT AND INCOHERENT PROCESSING

Consider a hypothetical optical system as depicted in Fig. 9.6. The light emitted by the source Σ is assumed monochromatic. To determine the light field at the output plane, we let this complex light distribution at the input plane, due to $d\Sigma$ be $u(x, y)$. If the complex amplitude transmittance of the input plane is $f(x, y)$, the complex light field immediately behind the signal plane would be $u(x, y) f(x, y)$. By assuming the optical system is linear and spatially invariant, the complex light field at the output plane of the system, due to $d\Sigma$, can then be written as

$$g(\alpha, \beta) = [u(x, y) f(x, y)] * h(x, y) \tag{9.17}$$

where the asterisk denotes the convolution operation, and the corresponding intensity distribution, due to Σ, is given by

$$I(\alpha, \beta) = \iint |g(\alpha, \beta)|^2 \, d\Sigma \tag{9.18}$$

which can be written as

$$I(\alpha, \beta) = \iiiint \Gamma(x, y; x', y') h(\alpha - x, \beta - y) h^*(\alpha - x', \beta - y') \\ \cdot f(x, y) f^*(x', y') \, dx \, dy \, dx' \, dy' \tag{9.19}$$

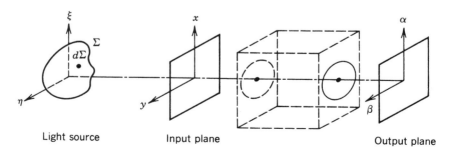

Fig. 9.6 A hypothetical optical signal-processing system.

where

$$\Gamma(x, y; x', y') = \iint_\Sigma u(x, y)u^*(x', y')\, d\Sigma \qquad (9.20)$$

is the *spatial coherence function*, which is also known as the *mutual intensity function*.

Let us now choose two arbitrary points Q_1 and Q_2 at the input plane. If r_1 and r_2 are the distances from Q_1 and Q_2 to $d\Sigma$, the respective complex light disturbances at Q_1 and Q_2, due to $d\Sigma$, can be written as

$$u_1(x, y) = \frac{[I(\xi, \eta)]^{1/2}}{r_1} e^{ikr_1} \qquad u_2(x', y') = \frac{[I(\xi, \eta)]^{1/2}}{r_2} e^{ikr_2} \qquad (9.21)$$

where $I(\xi, \eta)$ is the intensity distribution of the light source. By substituting Eqs. (9.21) in Eq. (9.20), we have

$$\Gamma(x, y; x', y') = \iint_\Sigma \frac{I(\xi, \eta)}{r_1 r_2} \exp[ik(r_1 - r_2)]\, d\Sigma \qquad (9.22)$$

When the light rays are paraxial, $r_1 - r_2$ can be approximated by

$$r_1 - r_2 \approx \frac{1}{r}[\xi(x - x') + \eta(y - y')] \qquad (9.23)$$

where r is the distance between the source and the signal plane. Then Eq. (9.23) becomes

$$\Gamma(x, y; x', y') = \frac{1}{r^2} \iint I(\xi, \eta) \exp\left\{i\frac{k}{r}[\xi(x - x') + \eta(y - y')]\right\} d\xi\, d\eta \qquad (9.24)$$

which is essentially the inverse Fourier transform of the source intensity distribution. Equation (9.24) is also known as the *Van Cittert-Zernike theorem*.

Let us now consider two extreme cases. In one we let the light source become infinitely large, that is, $I(\xi, \eta) \cong K$; Eq. (9.24) then becomes

$$\Gamma(x, y; x', y') = K_1 \delta(x - x', y - y') \qquad (9.25)$$

which describes a completely *incoherent* illumination, where K_1 is a proportionality constant.

On the other hand, if we let the light source be vanishingly small, that is, $I(\xi, \eta) \cong K\delta(\xi, \eta)$, Eq. (9.24) becomes

$$\Gamma(x, y; x', y') = K_2 \qquad (9.26)$$

which describes a completely *coherent* illumination, where K_2 is an arbitrary constant. In other words, a monochromatic point source describes a strictly coherent regime, whereas an extended source describes a strictly incoherent radiation.

For the completely incoherent illumination, the output intensity distribution can be shown as

$$I(\alpha, \beta) = \iint |h(\alpha - x, \beta - y)|^2 |f(x, y)|^2 \, dx \, dy \tag{9.27}$$

from which we see that incoherent optical system is *linear in intensity*.

On the other hand, under the strictly coherent illumination as described in Eq. (9.26), Eq. (9.19) becomes

$$\begin{aligned} I(\alpha, \beta) &= g(\alpha, \beta)g^*(\alpha, \beta) = \iint h(\alpha - x, \beta - y) f(x, y) \, dx \, dy \\ &\cdot \iint h^*(\alpha - x', \beta - y') f^*(x', y') \, dx' \, dy' \end{aligned} \tag{9.28}$$

from which we can see that the coherent optical system is *linear in complex amplitude*.

9.4 EXPLOITATION OF COHERENCE

In the preceding section, we saw that a coherent processor is capable of carrying out the complex amplitude processing, which offers a myriad of applications. However, coherent optical processing also suffers from coherent artifact noise, which limits its processing capabilities. To alleviate these limitations, we discuss methods to exploit the coherence contents from an incoherent source for complex amplitude (or coherent) processing. Since all physical sources are neither strictly coherent nor strictly incoherent, it is possible to extract the inherent coherence contents from an incoherent source for coherent processing.

9.4.1 Exploitation of Spatial Coherence

We begin our discussion on the exploitation of spatial coherence from an extended incoherent source. A conventional optical signal processor is shown in Fig. 9.7 in which an extended light source is used. With reference to the spatial coherence function of Eq. (9.24), we have

$$\Gamma(\mathbf{x}_2 - \mathbf{x}_2') = \iint \gamma(\mathbf{x}_1) \exp\left[i2\pi \frac{\mathbf{x}_1}{\lambda f}(\mathbf{x}_2 - \mathbf{x}_2')\right] d\mathbf{x}_1 \tag{9.29}$$

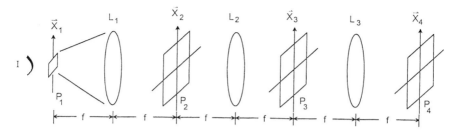

Fig. 9.7 An incoherent source optical processor: I, incoherent source; L_1, collimating lens; L_2 and L_3, achromatic transformation lenses; P_1, source encoding mask; P_2, input plane; P_3, Fourier plane; and P_4, output plane.

from this result, we see that the spatial coherence function and the source-encoding intensity form a Fourier transform pair, as given by

$$\gamma(\mathbf{x}_1) = \mathbf{F}[\Gamma(\mathbf{x}_2 - \mathbf{x}_2')]$$
$$\Gamma(\mathbf{x}_2 - \mathbf{x}_2') = \mathbf{F}^{-1}[\gamma(\mathbf{x}_1)] \tag{9.30}$$

where $\mathbf{F}$ denotes the Fourier transform operation. In other words, if a specific spatial coherence function is needed, the corresponding source-encoding function can be obtained by the Fourier transformation. Note that the source-encoding function $\gamma(x_1)$ can consist of apertures of different shapes or slits, but it should be a positive real function that satisfies the following physically realizable constraint:x

$$0 \leq \gamma(\mathbf{x}_1) \leq 1 \tag{9.31}$$

9.4.2 Exploitation of Temporal Coherence

Since the scale of the Fourier transform is linearly proportional to the wavelength of the light source, it is apparently not capable of (or inefficient at) using a broadband source. To do so, a narrow-spectral-band (i.e., temporally coherent) source is needed. In other words, the spectral spread of the input object should be confined within a small fringe spacing of the spatial filter, which is given by

$$\frac{p_m f \, \Delta\lambda}{2\pi} \ll d \tag{9.32}$$

where d is the fringe spacing of the spatial filter, p_m is the upper angular spatial frequency content of the input object, f is the focal length of the transform lens, and $\Delta\lambda$ is the spectral bandwidth of the light source. Then the spectral width (or the temporal coherence requirement) of the light source should satisfy the following constraint:

$$\frac{\Delta\lambda}{\lambda} \ll \frac{\pi}{h_0 p_m} \tag{9.33}$$

where λ is the center wavelength of the light source, $2h_0$ is the size of the input object transparency, and $2h_0 = (\lambda f)/d$.

There are, however, techniques to extract temporal coherence content from a broadband source. One of the simplest and most efficient method is by dispersing the Fourier spectra in the Fourier plane, which can be obtained by placing a spatial sampling grating at the input plane P_2. For example, if the input object is sampled by a phase grating as given by

$$f(x_2)T(x_2) = f(x_2) \exp(ip_0 x_2) \tag{9.34}$$

Then the corresponding Fourier transform would be

$$F(p, q) = F\left(x_3 - \frac{\lambda f}{2\pi} p_0\right) \tag{9.35}$$

in which we see that $F(p, q)$ is smeared into rainbow colors along the x_3 in the Fourier plane. Thus, a high degree of temporal coherence of the Fourier spectrum within a narrow-spectral-band filter can be obtained. In other words, by simply increasing the sampling frequency p_0 at the input domain P_2, a high degree of temporal coherent Fourier spectra can be obtained, as given by

$$\frac{\Delta\lambda}{\lambda} \cong \frac{4p_m}{p_0} \ll 1 \tag{9.36}$$

9.4.3 Complex Amplitude Processing

Let us describe a broad-spectral-band processor that exploits the coherent contents for complex amplitude processing. By referring to the incoherent-source processor of Fig. 9.7, an extended white light source is used. First let us look at the spatial coherence requirement, for which a source-encoding mask is used. We illustrate the complex signal detection, in which we assume that the target is much smaller than the entire input plan. Thus, the spatial coherence requirement should be imposed by the size of the target rather than the entire input spatial domain P_2. To ensure a physically realizable source-encoding mask, the required spatial coherence that covers the input target can be written as

$$\Gamma(|r_2 - r_2'|) = \frac{J_1[(\pi/h_0)|r_2 - r_2'|]}{(\pi/h_0)|r_2 - r_1'|} \tag{9.37}$$

where J_1 is a first-order Bessel function of first kind and h_0 is the size of the target. The corresponding spatial coherence as a function of $|r_2 - r_2'|$ is sketched in Fig. 9.8a, such that the source-encoding function is obtained as

$$\gamma(|r_1|) = \mathrm{cir}\left\{\frac{|r_1|}{w}\right\} \tag{9.38}$$

where $w = (f\lambda)/h_0$ represents the diameter of a circular aperture, f is the focal length of the collimating lens, and λ is the wavelength.

We now consider the next illustration for linear image deblurring. Since smeared image deblurring is a one-dimensional processing operation, inverse filtering takes place with respect to the smeared length of the blurred object. Thus, the required spatial coherence depends on the smeared length instead of the entire input plane. If we assume that a spatial coherence function is given by

$$\Gamma(x_2, x_2') = \mathrm{Sinc}\left\{\frac{\pi}{\Delta x_2}(x_2, x_2')\right\} \tag{9.39}$$

then the source-encoding function can be shown as

$$\gamma(x_1) = \mathrm{rect}\left(\frac{x_1}{w}\right) \tag{9.40}$$

where Δx_2 is the smeared length, $w = (f\lambda)/(\Delta x_2)$ is the slit width of the encoding aperture as shown in Fig. 9.8b, and

$$\mathrm{rect}\left(\frac{x_1}{w}\right) = \begin{cases} 1 & -\dfrac{w}{3} \leq x_1 \leq \dfrac{w}{2} \\ 0 & \text{otherwise} \end{cases} \tag{9.41}$$

We now consider the third example for image subtraction. Since the spatial coherence depends on the corresponding point-pair of the images to be subtracted, a strictly broad spatial coherence function is not required. In other words, a point-pair spatial coherence function is needed. To ensure the physically realizability of the source-encoding function, we assume the point-pair spatial coherence function is given by

$$\Gamma(x_2 - x_2') = \frac{\sin[(N\pi/h_0)(x_2 - x_2')]}{N \sin[(\pi/h_0)(x_2 - x_2')]} \mathrm{Sinc}\left[\frac{\pi w}{h_0 d}(x_2 - x_2')\right] \tag{9.42}$$

where $2\,h_0$ is the main separation of the two input image transparencies. As $N \gg 1$ and $w \ll d$, preceding equation represents a sequence of narrow pulses that occur at $(x_2 - x_2') = nh_0$, where n is a positive integer. Thus, we see that a high degree of coherence of the point-pair function can be obtained. By

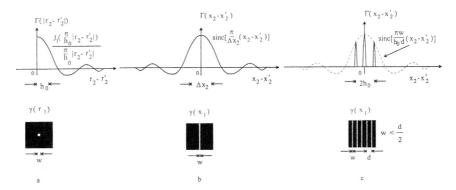

Fig. 9.8 Source encoding and spatial coherence; Γ, spatial coherence function, and γ, and source-encoding mask. (a) Correlation detection, (b) image deblurring, and (c) image subtraction.

Fourier transforming the preceding equation, the source-encoding function can be shown as

$$\gamma(x_1) = \sum_{n=1}^{N} \text{rect}\left(\frac{x_1 - nd}{w}\right) \qquad (9.43)$$

where w is the slit width, and $d = (\lambda f)/h_0$ is the separation between the slits. By plotting the preceding equation, the source-encoding mask is in fact represented by N equally spaced narrow slits, as shown in Fig. 9.8c.

We now investigate the temporal coherence requirement, for which a sampling grating is used to disperse the Fourier spectrum in the Fourier plane. Let us consider the temporal coherence requirement for two-dimensional complex signal detection. By referring to Eq. (9.36), a high degree of temporal coherence can be achieved by using a higher sampling frequency. We assume that the Fourier spectrum dispersion due to the sampling grating is along the x direction and the narrow-spectral-band matched filter is limited by the wavelength spread of Eq. (9.33), that is,

$$\Delta\lambda = \frac{\pi\lambda}{10h_0 p_m} \qquad (9.44)$$

Thus we see that the spatial width (or size) of the narrow-spectral-band filter is given by

$$\Delta w = \frac{f\lambda}{\pi} p_m \qquad (9.45)$$

In view of Eq. (9.36), the sampling frequency can be determined as

$$p_0 = \frac{4\lambda p_m}{\Delta \lambda} \tag{9.46}$$

In other words, the higher the sampling frequency used, the more efficient the processing performance would be. We see that a larger set of temporal matched filters can be used, which results in higher output image irradiance. However, the apparent price paid is using large achromatic transform lenses, which tend to be more expensive.

9.4.4 Incoherent Superposition

In the preceding section, we saw that spatial and temporal coherence can be exploited by spatial encoding and spectral dispersion of an incoherent source. We see that complex amplitude processing can be carried out with a set of narrow-spectral-band filters, as given by

$$H_n(p_n, q_n; \lambda_n) \quad \text{for } n = 1, 2, \ldots, N, \tag{9.47}$$

where (p_n, q_n) represents the angular frequency coordinates and λ_n is the center wavelength of the narrow-width filter. Then it can be shown that the output light intensity would be the incoherent superposition of the filtered signals, as given by

$$I(x, y) \cong \sum_{n=1}^{N} \Delta \lambda_n |f(x, y; \lambda_n) * h(x, y; \lambda_n)|^2 \tag{9.48}$$

where the asterisk denotes the convolution operation, $f(x, y; \lambda_n)$ represents the input signal illuminated by λ_n, $\Delta\lambda_n$ is the narrow spectral width of the narrow-spectral-band filter, and $h(x, y; \lambda_n)$ is the spatial impulse response of $H(p_n, q_n; \lambda_n)$, that is,

$$h(p_n, q_n; \lambda_n) = \mathbf{F}^{-1}[H(p_n, q_n; \lambda_n)] \tag{9.49}$$

In view of the preceding result, the incoherent source processor can indeed process the information in complex amplitude as a coherent processor. Since the output intensity distribution is the sum of mutually incoherent image irradiances, the annoying coherent artifact noise can be suppressed.

However, if the signal processing is a one-dimensional operation, then the processing can be carried out with a fan-shaped broadband filter to compensate the scale variation of the Fourier spectra. The output intensity distribution can be shown as

$$I(x, y) = \int_{\Delta \lambda} |f(x, y; \lambda) * h(x, y; \lambda)|^2 \, d\lambda \tag{9.50}$$

where the integral is over the entire spectral band of the light source. Again, we see that the output irradiance is obtained by incoherent superposition of the entire spectral-band light source, by which we can in principle get rid of the coherent artifact noise.

9.5 REMARKS

The merit of using coherent light for optical data processing, in contrast to using incoherent light, must be the complex amplitude processing. Thus the coherent processing system has a much larger information-processing capacity as compared with the incoherent processing system.

One of the major advantages of using the broadband incoherent source for data processing must be the elimination of coherent artifact noise; Dennis Gabor regarded it as the number 1 enemy in optical signal processing. Although the system under incoherent illumination in principle cannot process the information in complex amplitude as the coherent system does, temporal and spatial coherence contents from an incoherent source can be exploited for complex amplitude processing. In other words, to exploit the spatial coherence content at the input domain, one can spatially encode the light source. And to exploit a high temporal coherence content at the Fourier domain, one can disperse the Fourier spectral in rainbow color. The advantage of exploiting the coherence content from an incoherent source is that the information can be processed in complex amplitude as by a coherent processor, and at the same time the coherent artifact noise can be suppressed as processed with an incoherent processor.

REFERENCES

1. M. Born and E. Wolf, *Principle of Optics*, 2nd rev. ed., Pergamon Press, New York, 1964.
2. F. T. S. Yu and S. Jutamulia, *Optical Signal Processing, Computing and Neural Networks*, Wiley-Interscience, New York, 1992.
3. D. Gabor, "Laser speckle and its elimination," *IBM J. Res. Dev.*, vol. 14, 509–574 (1970).
4. F. T. S. Yu, *White Light Signal Processing*, Wiley-Interscience, New York, 1985.
5. S. T. Wu and F. T. S. Yu, "Image subtraction with encoded extended incoherent source," *Appl. Opt.*, vol. 20, 4082–4088 (1988).
6. F. T. S. Yu, "Exploitation of incoherent source for partially coherent processing," *Opt. Eng.*, vol. 36, 1458–1464 (1997).

10
Wavelet Transforms with Optics

Two major questions concerning this chapter may be asked: What is a wavelet? Why is it interesting for solving signal processing problem? These are crucial remarks that a signal analyst would want to know. The answer to these questions may be summarized as: Wavelet representation is a versatile technique having many physical and mathematical insights with great potential applications to signal and image processing. In other words, wavelets can be viewed as a new basis for representing signals and images that can be used for signal analysis and image synthesis.

10.1 ASPECTS OF WAVELET TRANSFORM

For nonstationary signal processing, the natural way to obtain joint time-frequency resolution of a signal is to take the Fourier transform of the signal within a time window function. This transform is called *short-time Fourier transform* (STFT), where the size of window is assumed invariant with respect to frequency content of the signal [1–4]. On the other hand, if the size of the window changes as the analyzing frequency changes, then the transform is known as *wavelet transform* (WT) [4–7].

The expression of the STFT can be written as [1–3]

$$\text{STFT}(\tau, \omega) = \int x(t) h^*(t - \tau) \exp(-i\omega t)\, dt \tag{10.1}$$

where $h(t)$ is an analyzing window function, ω is a selected analyzing frequency and τ is an arbitrary time shift. Notice that if $h(t)$ is a Gaussian function, then the transform is also known as the *Gabor transform*. The STFT has been widely used in signal processing, such as time-varying signal analysis and filtering, spectral estimation, signal compression, and others [1–4]. Usually the STFT offers very good performances for signals having uniform energy distribution within an analyzing window. Thus, the selection of the analyzing window size is critically important for achieving an optimum joint time-frequency resolution. However, an apparent drawback that prevents the STFT from wider application is that the invariant size of the analyzing window leads to a lower time resolution for nonuniformly distributed signals, such as transient, discontinuous signals. Therefore, to overcome this shortcoming, the WT can be used as given by [4–7]

$$\text{WT}(\tau, a) = \frac{1}{\sqrt{a}} \int x(t) \psi^* \left(\frac{t - \tau}{a} \right) dt \qquad (10.2)$$

where a is a scaling factor, $\psi(t)$ is called *mother wavelet*, and τ is an arbitrary time shift. Notice that the shape of the *mother wavelet* $\psi(t)$ shrinks as the scaling factor a decreases, while dilates as a increases. The shrunk and dilated wavelets are also known as the *daughter wavelets* [5, 6]. Thus, to have a better time resolution, a narrower WT window should be used for higher frequency content. In principle, the WT suffers from the same time-frequency resolution limitation as the STFT, that is, the time resolution and the frequency resolution cannot be resolved simultaneously as imposed by the inequity

$$\Delta t \, \Delta \omega \geq 2\pi \qquad (10.3)$$

where Δt and $\Delta \omega$ can be written as

$$\Delta t = \frac{\int |h(t)| \, dt}{|h(0)|} \quad \text{and} \quad \Delta \omega = \frac{\int |H(\omega)| \, d\omega}{|H(0)|}$$

in which $H(\omega)$ is the Fourier transform of $h(t)$. The product of Δt and $\Delta \omega$ is known as *logon*. In practice, window functions having smaller cells have better joint time-frequency localization. The *Gaussian window function* is the best one in the sense of meeting the lower bound of the inequality. However, the Gaussian function lacks either biorthogonality or orthogonality, which is the constraint of window functions for perfect signal reconstruction. Perfect reconstruction is one of the objectives of the STFT and the WT, such as nonlinear filtering, signal compression and synthesis.

10.2 FOURIER DOMAIN PROCESSING

We start from the basic definitions of the semicontinuous STFT and the semicontinuous WT for which general window functions are given [9]. There are two types of the STFT and the WT, namely, complex-valued types and real-valued types. To discuss the wavelet transform, we begin with the semicontinous STFT, defined as

$$\text{STFT}(\tau, n) = \frac{1}{\sqrt{2\pi}} \int x(t) h^*(t - \tau) \exp(-in\omega_0 t) \, dt \qquad (10.4)$$

where n is an integer, $n\omega_0$ is the selected analyzing frequency, and τ is an arbitrary time shift. In Fourier domain representation, Eq. (10.4) can be written as

$$\text{STFT}(\tau, n) = \frac{1}{\sqrt{2\pi}} \exp(-in\omega_0 \tau) \int X(\omega) H^*(\omega - n\omega_0) \exp(i\omega \tau) \, d\omega \tag{10.5}$$

where $X(\omega)$ and $H(\omega)$ are the Fourier transforms of $x(t)$ and $h(t)$, respectively, and the super arterial asterisk denotes the complex conjugate. Since $h(t)(-in\omega_0 t)$ can be regarded as a bandpass filter centered at $n\omega_0$, $\text{STFT}(\tau, n)$ is actually a multifilter system in which the bandwidth of the filters are invariant in the entire frequency domain. Although any smooth function with finite energy can be used as an analyzing window, windows having lower uncertainty are preferred. On the other hand, in nonlinear filtering, the perfect-reconstruction of a signal is the main issue, such that the window function $h(t)$ is subjected to some physical constraints.

The inverse operation of the STFT can be written as

$$x(t) = \frac{1}{\sqrt{2\pi}} \sum_{n=-\infty}^{+\infty} \int \text{STFT}(\tau, n) s(t - \tau) \exp(in\omega_0 t) \, d\tau \tag{10.6}$$

which can also be expressed in terms of frequency domain, such as

$$x(t) = \frac{1}{\sqrt{2\pi}} \sum_{n} \int \mathcal{F}\{\text{STFT}(\tau, n)\}_\tau S(\omega - n\omega_0) \exp(i\omega t) \, d\omega \tag{10.7}$$

where $s(t)$ is a given synthesis window function, its Fourier transform is given as $S(\omega)$, and $\mathcal{F}\{\ \}_\tau$ represents the Fourier transform with respect to τ. For a perfect reconstruction, the relationship between $h(t)$ and $s(t)$ can be formed by the following biorthogonality constraint [3,4]:

$$\sum_{n=-\infty}^{+\infty} H^*(\omega - n\omega_0) S(\omega - n\omega_0) = C \tag{10.8}$$

where C is a constant which need not be unity.

It is well known that $s(t)$ cannot be uniquely determined and for which there are infinitely choices that satisfy the constraint of Eq. (10.8). A logical choice for a symmetrical forward and inverse operation as described in Eq. (10.4) and Eq. (10.6) is that $s(t) = h(t)$, for which $h(t)$ is called an orthogonal function. For the other choices of $s(t) \neq h(t)$, $h(t)$, and $s(t)$ are called *biorthogonal functions*. There is, however, an attractive biorthogonal pair of $h(t)$ and $\delta(t)$ [i.e., $s(t) = \delta(t)$] very suitable for the optical implementation of the STFT since it results in a simple inverse operation, where $\delta(t)$ is a Dirac delta function.

We notice that by choosing $s(t) = \delta(t)$ or $s(t) = h(t)$, Eq. (10.8) can be written in terms of the biorthogonality and the orthogonality constraints as follows:

$$\sum_{n=-\infty}^{+\infty} H^*(\omega - n\omega_0) = C \quad \text{where } s(t) = \delta(t) \quad (10.9\text{a})$$

$$\sum_{n=-\infty}^{+\infty} |H(\omega - n\omega_0)|^2 = C \quad \text{where } s(t) = h(t) \quad (10.9\text{b})$$

Although the Gaussian function has been shown to be one of the best window functions maximally concentrated in time and frequency, it cannot perfectly reconstruct a signal as referred to the constraints of Eq. (10.9), that is, the Gaussian function lacks the biorthogonality. A simple window $h(t)$ that satisfies the constraints of Eq. (10.9) is a Sinc function for which its Fourier transform is given by

$$H(\omega) = \text{rect}\left(\frac{\omega}{\omega_0}\right) = \begin{cases} 1 & |\omega| \leq \omega_0 \\ 0 & |\omega| > \omega_0 \end{cases}$$

Nevertheless, the high order sidelobes of a Sinc window function vanish rather slowly as a function of t. By referring to the constraints of Eq. (10.9a), a general form of $H(\omega)$ can be constructed as given by

$$H(\omega) = \begin{cases} \sin^2\left[\frac{\pi}{2}v\left(\frac{\omega_1 + \omega}{\omega_1 - \omega_2}\right)\right] & -\omega_2 \leq \omega \leq -\omega_1 \\ 1 & -\omega_1 \leq \omega \leq \omega_1 \\ \cos^2\left[\frac{\pi}{2}v\left(\frac{\omega - \omega_1}{\omega_2 - \omega_1}\right)\right] & \omega_1 \leq \omega \leq \omega_1 \\ 0 & |\omega| > \omega_2 \end{cases} \quad (10.10)$$

which forms a biorthogonal window function in time domain, where v is a function that satisfies $v(x) + v(1 - x) = 1$, $0 \leq x \leq 1$. An example of $v(x)$, as shown in Fig. 10.1, is given as

$$v_n(x) = \sin^2\left[\frac{\pi}{2}v_{n-1}(x)\right] \quad (10.11)$$

where $v_0(x) = x$, and $n > 1$ is an integer. We see that as $n \to \infty$, $v_n(x)$ approaches to a step function which will produce a rectangular $H(\omega)$. We would mention that a square-root version of Eq. (10.10) satisfies the constraint of Eq. (10.9b) which in turn produces an orthogonal window function.

Wavelet Transforms with Optics

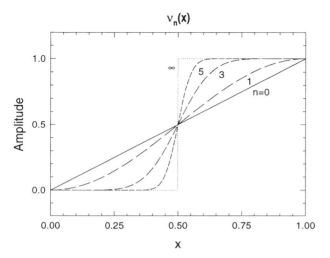

Fig. 10.1 v function as n increases.

Referring to the variations of $v(x)$ in Fig. 10.1, the Fourier domain representation of a set of typical window functions $h(t)$ is shown in Fig. 10.2a, as given by

$$H_1(\omega) = \text{rect}(2\omega)$$

$$H_2(\omega) = \begin{cases} \cos\left(\frac{\pi}{2}\omega\right) & |\omega| \leq 1 \\ 0 & \text{others} \end{cases}$$

$$H_3(\omega) = \begin{cases} \cos^2\left(\frac{\pi}{2}\omega\right) & |\omega| \leq 1 \\ 0 & \text{others} \end{cases}$$

$$H_4(\omega) = \exp(-\pi\omega^2)$$

The corresponding time domain windows are

$$h_1(t) = \frac{\sin(t/2)}{t/2}$$

$$h_2(t) = \frac{\pi \cos(t)}{(\pi/2)^2 - t^2}$$

$$h_3(t) = \frac{\pi^2 \sin(t)}{t(\pi^2 - t^2)}$$

$$h_4(t) = \exp\left(-\frac{t^2}{4\pi}\right)$$

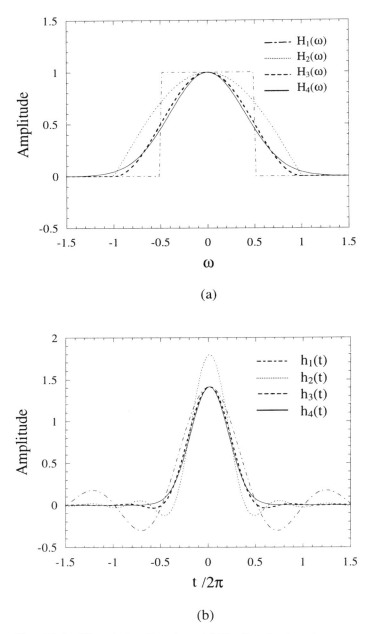

Fig. 10.2 The window functions. (a) The Fourier domain representations, $H_1(\omega)$, $H_2(\omega)$, $H_3(\omega)$, $H_4(\omega)$; (b) the windows $h_1(t)$, $h_2(t)$, $h_3(t)$, $h_4(t)$.

In view of the frequency-time transformation relationship shown in Fig. 10.2, we see that the window $h_3(t)$ exhibits a close resemblance to the Gaussian window $h_4(t)$, in addition to the Fourier domain [i.e., $H_3(\omega) \sim H_4(\omega)$]. One of the merits of using $h_3(t)$ is that it is a biorthogonal function which can simplify the inverse operation since the signal synthesis is just the summation of all the STFT signals from each analyzing filter band, as can be seen from Eq. (10.9a). Notice that the Sinc window $h_1(t)$ shows the disadvantage of slow vanishing sidelobes, but it is a practical filter of rectangular form in the Fourier domain. One of the interesting properties of the $h_1(t)$ is having both orthogonality and biorthogonality in terms of Eqs. (10.9a) and (10.9b). The window $h_2(t)$ is orthogonal in terms of the constraint of Eq. (10.9b) and has comparable joint time-frequency resolution with the window $h_3(t)$.

A quantitative comparison among these window functions in terms of joint time-frequency resolution cell is tabulated in Table 10.1. We can see that the cell (i.e., $\Delta t\, \Delta\omega$) of the biorthogonal window $h_3(t)$ almost meets the lower bound of the inequity equation (10.3). The resolution cell of the orthogonal window $h_2(t)$ is also close to the lower bound of the uncertainty principle, but almost one and half times larger than that of $h_3(t)$. On the other hand, because of its ideal cutoff frequency, the window $h_1(t)$ has the largest cell which could be regarded as the upper bound of the joint time-frequency resolution cell for nontrivial smooth window functions.

One of the objective is to design biorthogonal windows which would provide a smaller cell of joint time-frequency resolution. In view of Fig. 10.2, the biorthogonal window $h_3(t)$ happens to be an excellent approximation to the Gaussian window and almost meets the lower bound of the uncertainty inequity. Furthermore, an accompanying advantage of the window $h_3(t)$ is the simplification of the inverse STFT.

10.3 WAVELET TRANSFORM

A semicontinuous WT can be written as

$$\mathrm{WT}(\tau, n) = a_0^n \int x(t)\psi^*[a_0^n(t-\tau)]\, dt \tag{10.12}$$

and its Fourier domain representation is

$$\mathrm{WT}(\tau, n) = \frac{1}{2\pi} \int X(\omega)\Psi^*(a_0^{-n}\omega)\exp(i\omega\tau)\, d\omega \tag{10.13}$$

Table 10.1 Joint Time-Frequency Resolution Cell $\Delta t\, \Delta \omega$ of the Window Functions

Resolution Cell	Windows			
	$h_1(t)$	$h_2(t)$	$h_3(t)$	$h_4(t)$
$\Delta \omega$	1.0	1.27	1.0	1.0
Δt	20.28	7.40	6.54	2π
$\Delta t\, \Delta \omega$	20.28	9.43	6.54	6.28

where $a_0 > 1$ is a scaling factor (i.e., $a = a_0^{-n}$), n is an integer, $\psi(t)$ is the mother wavelet, and $\Psi(\omega)$ is its Fourier transform. Equation (10.12) is somewhat different from Eq. (10.2), where the normalization factor $1/\sqrt{a}$ (i.e., $a_0^{n/2}$) is used instead of a_0^n (i.e., $1/a$). We notice that the modification of Eq. (10.12) would simplify the optical implementation of the WT in the FDP. As has been pointed by Telfer and Szu [10] the frequency bias problem can be in fact alleviated by using the normalizing scaling factor $1/a$. Similar to the STFT, the WT can also be regarded as a multifilter system in the Fourier domain by which a signal can be decomposed into different spectral bands. Although the WT uses narrower and wider band filters to analyzing the lower- and the higher-frequency components, respectively, the operation is essentially similar to that of the STFT.

To meet the admissibility condition of the WT, $\psi(t)$ has to be a bandpass filter; however, for signals having rich low-frequency components, a scaling function $\phi(t)$ to preserve the low-frequency contents is needed for which the scaling transform of the signal is given by

$$\mathrm{ST}(\tau) = \int x(t) \phi^*(t - \tau)\, dt \tag{10.14}$$

Thus, the inverse operation of the WT can be written as

$$x(t) = \int \mathrm{ST}(\tau) s_\phi(t - \tau)\, d\tau + \sum_{n=-\infty}^{\infty} \int \mathrm{WT}(\tau, n) s_\psi[a_0^n(t - \tau)] a_0^n\, d\tau \tag{10.15}$$

and its Fourier domain representation is given by

$$\begin{aligned} x(t) = &\frac{1}{2\pi} \int \mathcal{F}\{\mathrm{ST}(\tau)\} S_\phi(\omega) \exp(i\omega t)\, d\omega \\ &+ \frac{1}{2\pi} \sum_{n=-\infty}^{\infty} \int \mathcal{F}\{\mathrm{WT}(\tau, n)\}_\tau S_\psi(a_0^{-n}\omega) \exp(i\omega t)\, d\omega \end{aligned} \tag{10.16}$$

where $s_\phi(t)$ and $s_\psi(t)$ are the synthesis scaling function and the wavelet function, respectively, and $S_\phi(\omega)$ and $S_\psi(\omega)$ are the corresponding Fourier transforms. If the WT is used for signal synthesis, then, for a perfect reconstruction, $\phi(t)$ and $\psi(t)$ must satisfy the following biorthgonality and orthogonality constraints:

$$\Phi^*(\omega) + \sum_{n=-\infty}^{+\infty} \Psi^*(a_0^{-n}\omega) = C \quad \text{for } s_\phi(t) = \delta(t) \quad \text{and} \quad s_\psi(t) = \delta(t)$$

(10.17a)

$$|\Phi(\omega)|^2 + \sum_{n=-\infty}^{+\infty} |\Psi(a_0^{-n}\omega)|^2 = C \quad \text{for } s_\phi(t) = \phi(t) \quad \text{and} \quad s_\psi(t) = \psi(t)$$

(10.17b)

Similar to Eq. (10.10) for the STFT, a general form of $\Psi(\omega)$ that satisfies the biorthogonality constraints of Eq. (10.17a) is given by

$$\Psi(\omega) = \begin{cases} 0 & \omega < \omega_0 \\ \sin^2\left[\frac{\pi}{2}v\left(\frac{\omega - \omega_0}{\omega_0(a_0 - 1)}\right)\right] & \omega_0 \leq \omega \leq a_0\omega_0 \\ \cos^2\left[\frac{\pi}{2}v\left(\frac{\omega - a_0\omega_0}{\omega_0 a_0(a_0 - 1)}\right)\right] & a_0\omega_0 \leq \omega \leq a_0^2\omega_0 \\ 0 & \omega > a_0^2\omega_0 \end{cases}$$

(10.18)

for which the scaling function $\Phi(\omega)$ can be shown as

$$\Phi(\omega) = \cos^2\left[\frac{\phi}{2}v\left(\frac{|\omega|}{(a_0 + 1)\omega_0/2}\right)\right] \qquad |\omega| \leq (a_0 + 1)\omega_0/2 \qquad (10.19)$$

where the function $v(x)$ has the same definition for the STFT as shown in Fig. 10.1. Thus, we see that $\Psi(\omega)$ forms a biorthogonal wavelet, and its squared-root version $\sqrt{\Psi(\omega)}$ is known as *Mayer's wavelet* [21] which is orthogonal in terms of the constaint of Eq. (10.17b).

Figure 10.3a shows the Fourier domain representations of a set of wavelets similar to the windows shown in Fig. 10.2, as given by

$$\Psi_1(\omega) = \begin{cases} 1 & 1.5 \leq \omega \leq 3 \\ 0 & \text{others} \end{cases}$$

$$\Psi_2(\omega) = \begin{cases} \sin[(\omega-1)\pi/2] & 1 \leq \omega \leq 2 \\ \cos[(\omega-2)\pi/4] & 2 \leq \omega \leq 4 \\ 0 & \text{others} \end{cases}$$

$$\Psi_3(\omega) = \begin{cases} \sin^2[(\omega-1)\pi/2] & 1 \leq \omega \leq 2 \\ \cos^2[(\omega-2)\pi/4] & 2 \leq \omega \leq 4 \\ 0 & \text{others} \end{cases}$$

$$\Psi_4(\omega) = \begin{cases} \exp[-\pi(\omega-2)^2] & \omega \leq 2 \\ \exp[-\pi(\omega-2)^2/4] & \omega \geq 2 \end{cases}$$

where the scaling factor a_0 is assumed 2. The real parts of the complex-valued $\psi_1(t)$, $\psi_2(t)$, $\psi_3(t)$, $\psi_4(t)$ are plotted in Figs. 10.3b, c, d, respectively, in which we see that the biorthogonal window $\psi_3(t)$ is an excellent approximation to the Gaussian function $\psi_4(t)$ both in the Fourier and the time domains. Therefore, the wavelet $\psi_3(t)$ has the advantages of having smaller joint time-frequency resolution cell and biorthogonality which can simplify the inverse operation of the WT. Also, $\psi_2(t)$ can be used as an orthogonal wavelet which has good joint time-frequency resolution. However, the wavelet $\psi_1(t)$ is often to use in practice since its Fourier transform is a rectangular form, which is rather convenient for the Fourier domain processing in making the masks of spatial filter banks.

Although the preceding discussions were limited to the wavelets which have similar forms to the window functions as discussed for the STFT, in practice, the WT has more solutions than the STFT. Namely, one can select particular wavelets for specific applications.

10.4 OPTICAL IMPLEMENTATIONS

Figure 10.4a shows an optical setup for the implementation of the STFT and the WT for one-dimensional signal processing [11], in which we assume that the synthesis function $s(t) = \delta(t)$. Thus, one should use biorthogonal windows and wavelets for the STFT and the WT, respectively, for perfect reconstruction. First, an input signal is displayed on a spatial light modulator (SLM) which is placed at the front focal plane of a one-dimensional Fourier transform lens CL1; then a set of filter banks are placed at the Fourier plane P2. Thus, the STFT (or the WT) signals

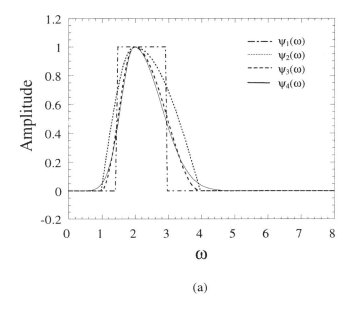

(a)

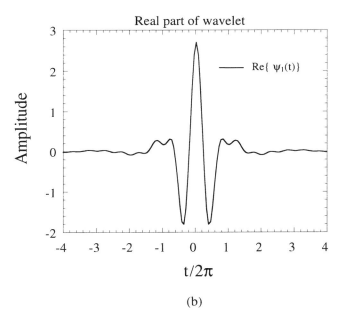

(b)

Fig. 10.3 (a) Fourier domain representations, $\Psi_1(\omega)$, $\Psi_2(\omega)$, $\Psi_3(\omega)$, $\Psi_4(\omega)$, of the complex-valued wavelets; (b) $\psi_1(t)$; (c) $\psi_2(t)$; (d) $\psi_3(t)$ and $\psi_4(t)$.

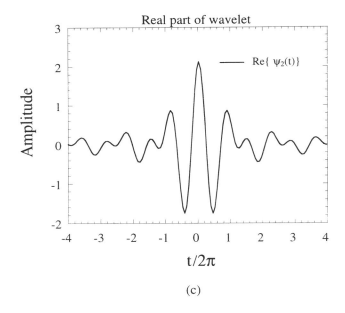

(c)

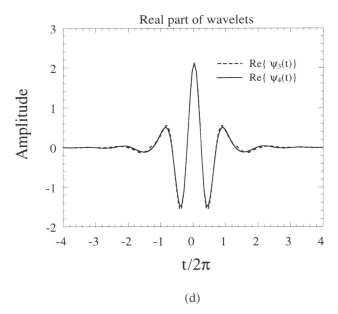

(d)

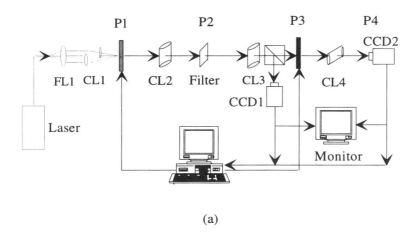

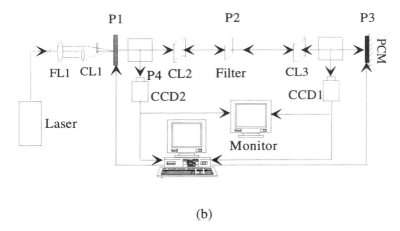

Fig. 10.4 Optical architectures for implementations of the STFT and the WT: (a) For biorthogonal windows or wavelets [i.e., $s(t) = \delta(t)$]; (b) for orthogonal windows or wavelets [i.e., $s(t) = h(t)$ or $\psi(t)$].

can be obtained in the back focal plane of the CL3 by capturing the spectrogram, i.e., intensity of the STFT with the CCD1. The reconstructed signal can be obtained at the output plane P4 by summing all he STFT (or the WT) signals diffracted from the filter banks. We notice that a real-time signal filtering can be realized by simply inserting a filter at the plane P3. Let us assume that the desired filter is $F(\tau, n)$, then the reconstructed signal can be shown as

$$x'(t) = \begin{cases} \sum_n \text{STFT}(\tau, n) F(\tau, n) & \text{for STFT} \\ \text{ST}(\tau) + \sum_n \text{WT}(\tau, n) F(\tau, n) & \text{for WT} \end{cases} \quad (20.20)$$

Figure 10.4b shows an optical setup in which we assume that the orthogonal windows (i.e., $s(t) = h(t)$) [or orthogonal wavelets, i.e., $s(t) = \psi(t)$] are used. Notice that the optical configuration of Fig. 10.4b is very similar to that of Fig. 10.4a, except for the inverse operation. By virtue of the reciprocity principle, the inverse operation can be accomplished by placing a phase conjugate mirror (PCM) behind the plane P3. The major advantage of the PCM is the self-alignment; therefore the filter alignment can be avoided. As the return phase conjugate signal STFT* (τ, n) [or WT* (τ, n)] is arrived at the plane P2, it will be subjected to the Fourier transformation. By inserting a desired filter at the plane P3, a real-time signal filtering can be obtained. We note that the filter at the plane P3 is proportional to $[F(\tau, n)]^{1/2}$, since the signal has to go through the filter twice. Thus the reconstructed signal at the plane P4 will be

$$x_r^*(t) = \begin{cases} \sum_n \int \mathcal{F}\{\text{STFT}^*(\tau, n) F(\tau, n)\}_\tau H^*(\omega - n\omega_0) \exp(i\omega t) \, d\omega & \text{for STFT} \\ \int \mathcal{F}\{\text{ST}^*(\tau) F(\tau)\}_\tau \Phi^*(\omega) \exp(i\omega t) \, d\omega \\ + \sum_n \int \mathcal{F}\{\text{WT}^*(\tau, n) F(\tau, n)\}_\tau \Psi^*(a_0^n \omega) \exp(i\omega t) \, d\omega & \text{for WT} \end{cases}$$

which can be observed with a CCD camera.

Although our preceding discussions are limited to one-dimensional signal processing, the basic concepts can be extended to two-dimensional image processing. This goal can be accomplished by either changing the alignment of filter banks or using photorefractive crystal filters and photorefractive crystal phase conjugate mirrors.

10.5 SIMULATIONS

Simulations of Fourier domain filter banks for complex-valued STFT and WT are shown in Fig. 10.5a and b, respectively. The filters centered at different frequencies are arranged from top to bottom. We see that the bandwidths of the filters are the same for the STFT, but vary in terms of the center frequencies for the WT (known as constant Q-filters). We have

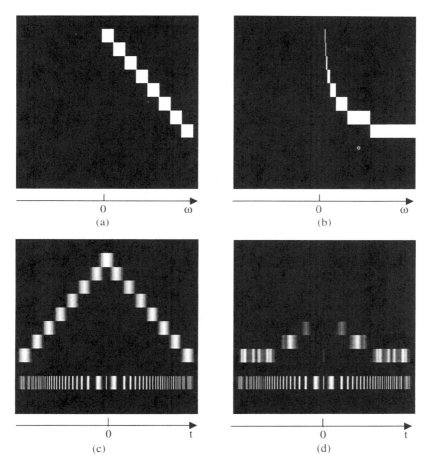

Fig. 10.5 Computer simulations of the complex-valued STFT and the complex-valued WT of a chirp with a transient signal at the center: (a) filter banks for the complex-valued STFT; (b) filter banks for the complex-valued WT; (c) the spectrogram of the test signal shown at the bottom; (d) the scalegram.

omitted the filters for negative frequency components since the spectrogram and scalegram from the negative frequency components is the same as that from the positive components. A test signal that includes a chirp and a transient signal is shown at the bottom of Fig. 10.5c and d, as given by

$$x(t) = \sin\left[\tfrac{\pi}{256}(t-127)^2\right] + \cos\left[\tfrac{3\pi}{4}(t-127)\right]\exp\left[-\tfrac{|(t-127)|}{2}\right]$$
$$0 \le t < 256$$

The corresponding spectrogram and the scalegram are shown at the top portion of Figs. 10.5c and d, in which we see that the STFT gives a relatively good time-frequency resolution for the chirp signal and a weaker response for the transient signal (located at the center). On the other hand, the WT provides a higher time resolution for the transient signal, but relatively poorer performance for the uniformly distributed chirp signal.

Real-valued STFT and WT are also simulated, and their filter banks are shown in Fig. 10.6a and b, respectively. Figure 10.6c and d show

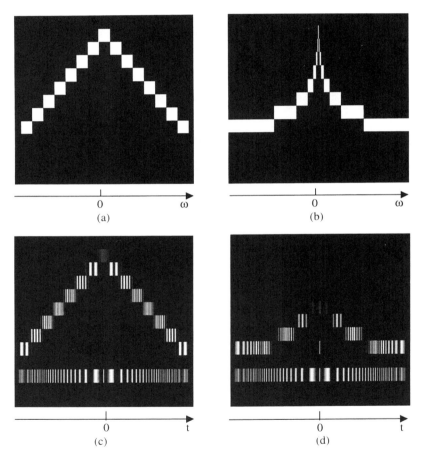

Fig. 10.6 Computer simulations of the real-valued STFT and the real-valued WT of a chirp with a transient signal at the center: (a) filter banks for the real-valued STFT; (b) filter banks for the real-valued WT; (c) the spectrogram of the test signal shown at the bottom; (d) the scalegram.

the spectrogram and the scalegram using the same test signal as in the complex-valued case. In view of these results, we see that the real-valued STFT and the real-valued WT provide the same time resolution as those of the complex-valued STFT and the complex-valued WT, respectively. However, the spectrogram and the scalegram are modulated by interference fringes, which would sometimes cause the loss of some details and difficulty of interpretations.

10.6 REMARKS

We have discussed the semicontinuous STFT, the semicontinuous WT, and their optical implementation. We have shown that one of the best substitutions for the Gaussian function in the Fourier domain is a squared sinusoid function which can form a biorthogonal window function in the time domain. A couple of optical architectures for the STFT and the WT are illustrated. Although our discussions are confined to one-dimensional signal processing, the basic concepts can be easily extended to multidimensional signal processing.

REFERENCES

1. D. Gabor, "Theory of Communication," *J. Inst. Elec. Eng.*, vol. 93, pp. 429–457, 1946.
2. L. Cohen, "Time-Frequency Distribution—A Review," *Proc IEEE*, vol. 77, pp. 941–981, 1989.
3. F. Hlawatsch and G. F. Boudreaux-Bartels, "Linear and Quadratic Time-Frequency Signal Representations," *IEEE SP Magazine*, pp. 21–67, April 1992.
4. A. N. Akansu and R. A. Haddad, *Multiresolution Signal Decomposition*. Academic Press, Inc., New York 1992, chap. 5, p. 291.
5. R. A. Haddad, A. N. Akansu, and A. Benyassine, "Time-Frequency Localization in Transforms, Subbands, and Wavelets: A Critical Review," *Opt. Eng.*, vol. 32, pp. 1411–1429, 1993.
6. O. Rioul and M. Vetterli, "Wavelets and Signal Processing," *IEEE SP*, pp. 14–38, October 1991.
7. C. H. Chui, *Wavelets: A Tutorial in Theory and Applications*. Academic Press, Inc., New York, 1992, chap. 1, p. 3.
8. M. B. Ruskai, G. Beylkin, R. Coifman, I. Daubchies, S Mallat, Y. Meyer, and L. Raphael, *Wavelets and their Applications*. Jones and Bartelett Publishers, Boston, 1992, chap. IV, p. 241.
9. I. Daubechies, "The Wavelet Transform, Time-Frequency Localization and Signal Analysis," *IEEE Trans. Information Theory*, vol. 36, pp. 961–1005, 1990.

10. B. Telfer and H. Szu, "New Wavelet Transform Normalization to Remove Frequency Bias," *Opt. Eng.*, vol. 31, pp. 1830–1834, 1992.
11. F. T. S. Yu and G. Lu, "Short-time Fourier transform (STFT) and Wavelet transform (WT) Using Fourier Domain Processing," *Applied Optics*, vol. 33, pp. 5262–5270, 1994.

11
Pattern Recognition with Optics

The roots of optical pattern recognition can be traced back to Abbe's work in 1873 [1], which led to the discovery of spatial filtering. However, optical pattern recognition was not appreciated until the complex spatial filtering of Vander Lugt in 1964 [2]. Since then, techniques, architectures, and algorithms have been developed to construct efficient optical correlators for pattern recognition. The objective of this chapter is to discuss the optical architecture and techniques as applied to pattern recognition. Basically, there are two approaches in the optical implementation, namely, the correlation approach and the neural net approach. In the correlation approach, there are two frequently used architectures, the Vander Lugt correlator (VLC) and the joint transform correlator (JTC).

11.1 OPTICAL CORRELATORS

Optical pattern recognition can be accomplished by either Fourier domain or spatial domain filtering. Correlators that use Fourier domain filtering are known as VLCs, and spatial domain filtering is frequently used (JTC). The basic distinction between them is that VLC depends on a Fourier domain filter, whereas JTC depends on a spatial domain filter. In other words, the complex spatial detection of Vander Lugt is input scene *independent*, whereas the joint transform is input scene *dependent* [3].

It is apparent that a pure optical correlator has drawbacks that make certain tasks difficult or impossible to perform. The first problem is that optical systems are difficult to program, in the sense of a general-purpose digital computer. A second problem is that accuracy is difficult to achieve in Fourier optics. A third problem is that optical systems cannot easily be used to make decisions. Even simple decision making cannot be performed optically without the intervention of electronics. However, many optical deficiencies happen to be strong points in the electronic counterpart.

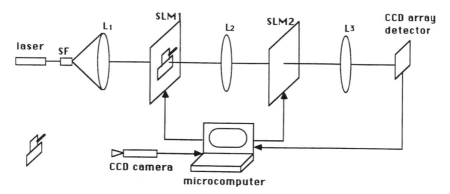

Fig. 11.1 Hybrid optical VLC. SF, spatial filter; L's lenses.

For example, accuracy, controllability, and programmability are obvious traits of digital computers. Thus, combining optical systems with electronic counterparts often produce better processors, as shown in Figs. 11.1 and 11.2, in which spatial light modulators (SLMs) are used for object and spatial filter devices. One of the important aspects of these hybrid-optical architectures is that decision making can be done by the computer.

To illustrate the adaptivity of the preceding JTC hybrid optical correlator, we provide an interesting application to perform autonomous target tracking. The idea is to correlate the target in the current frame with the target in the previous frame. This makes the hybrid JTC adaptive by constantly updating the reference target with the dynamic input scene. For example, two sequential scenes of a moving object are displayed on the SLM of Fig. 11.3 with the previous and the current frames positioned in the upper and lower half of the SLM, as given by

$$f_{t-1}\left(x - x_{t-1}, y - y_{t-1} - \frac{\alpha}{2}\right) \quad \text{and} \quad f_t\left(x - x_{t-1} - \delta x, y - y_{t-1} - \delta y + \frac{\alpha}{2}\right) \tag{11.1}$$

where 2α is the height of the display unit, t and $t-1$ represent the current and previous time frames, and $(\delta x, \delta y)$ is the relative translation of the object. Then the complex light field at the Fourier domain can be shown to be

$$T(u, v) = F_{t-1}(u, v) \exp\left\{-i2\pi\left[ux_{t-1} + v\left(y_{t-1} + \frac{\alpha}{2}\right)\right]\right\}$$
$$+ F_t(u, v) \exp\left\{-i2\pi\left[u(x_{t-1} + \delta x) + v\left(y_{t-1} + \delta y - \frac{\alpha}{2}\right)\right]\right\} \tag{11.2}$$

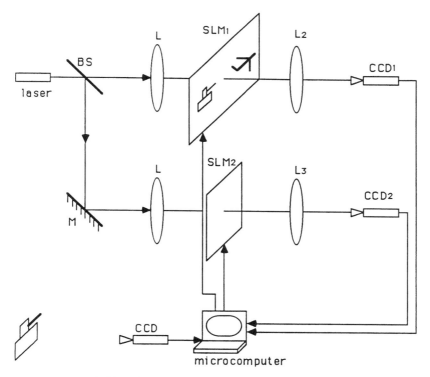

Fig. 11.2 Hybrid optical JTC. BS, beam splitter; M, mirror; L's, lenses.

If the detected $|T(u, v)|^2$ is sent back to the SLM, the output light distribution can be written as

$$C(x, y) = \mathcal{F}^{-1}\{|T(u, v)|^2\} = R_{t,t}(x, y) + R_{t-1,t-1}(x, y) \\ + R_{t,t-1}(x - \delta x, y + \delta y - \alpha) + R_{t-1,t}(x - \delta x, y - \delta y + \alpha) \quad (11.3)$$

where $R_{m,n}(x, y) = \mathcal{F}^{-1}\{F_m(u, v)F_n^*(u, v)\} = \iint f_m(u, v) f_n^*(n - x, v - y)\, du\, dv$ represents the correlations of f_m and f_n, which are diffracted at $x_1 = \delta x$, $y_1 = \delta y - \alpha$ and $x_2 = \delta x$, $y_2 = -\delta y + \alpha$, respectively. If the angular and scale tolerances of the JTC are approximately $\pm 5°$ and $\pm 10\%$ [4], and the motion of the target is relatively slow as compared to the processing cycle of the correlator, then f_{t-1} correlates strongly with f_t. Two high-intensity correlation peaks are diffracted into the output plane, to locations given by

$$x_t = x_{t-1} + x_1 \qquad y_t = y_{t-1} + y_1 + \alpha \quad (11.4)$$

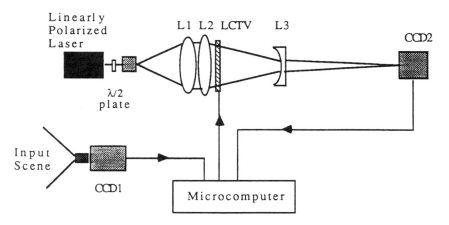

Fig. 11.3 A single SLM JTC. L's, lenses; LCTV, liquid-crystal TV.

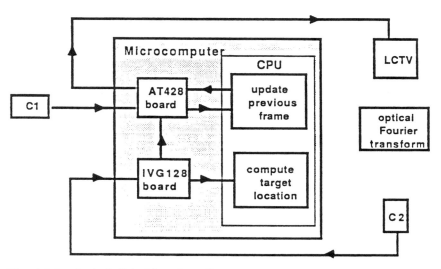

Fig. 11.4 Optical-digital interface diagram.

We note that a simple C language program can then be used to evaluate the target position, for which a block diagram of the system configuration is depicted in Fig. 11.4, for reference.

A major benefit of this technique is its adaptivity. For example, a situation where a camera mounted on a moving space vehicle is focused on a fixed target on the ground for automatic navigation. As the space vehicle

Pattern Recognition with Optics 255

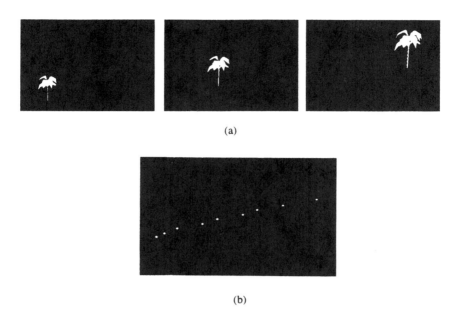

Fig. 11.5 (a) Sequence of nine images are recorded simulating the exact scenes as captured by a camera mounted on a moving space vehicle, with images 1, 5, and 9 shown here. (b) Tracked positions of the ground target as seen from the vehicle's coordinate frame.

approaches the target, the detected scene changes continuously: The target size appears larger and its orientation and shape change due to the motion of the vehicle. Using computer-aided design graphics, a three-dimensional treelike model was created as a simulated target on the ground. Nine image sequences simulating the changing scene viewed by a moving space vehicle are shown in part in Fig. 11.5a. The JTC tracking system has little difficulty in correlating targets from different frames, even though the target in the first and the last frame look different. Figure 11.5b shows the tracked locations of the target as seen from the vehicle.

11.2 OPTICAL-DISK-BASED CORRELATOR

Recent improvements in real-time addressable SLMs have made pattern recognition with optics much more realistic. A single-input image, however, might require correlation with a huge library of reference images. The lack of large-capacity information storage devices, which provide high-speed readout, is still a major obstacle to practical optical correlators. Optical

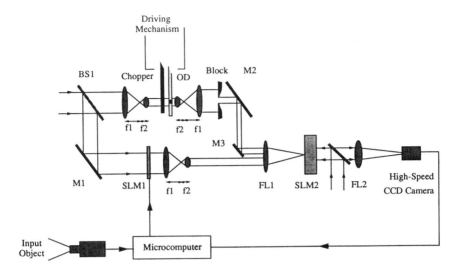

Fig. 11.6 An OD-based JTC. BS, beam splitter; M's, mirrors; FL's, Fourier-transform.

disks (ODs), developed in recent years as mass-storage media, are excellent candidates for this task. An OD-based JTC is shown in Fig. 11.6, in which a target captured by a video camera is displayed on SLM1. A beam expansion/reduction system is then used to reduce the size of the input image m times. The reference image stored on the optical disk is read out in parallel and magnified m times by another set of expansion lenses so that the reference image and input target are the same size. The joint transform is done by transform lens FL1, and the joint transform power spectrum (JTPS) is recorded on the write side of optically addressed SLM2. After FL2 Fourier transforms the JTPS, the resulting correlation is recorded by a high-speed CCD camera located at the output plane. The OD advances to the next reference image and so on until a match is found.

A transparent, write-once OD is used in this architecture [5]. To record a library of images, a laser beam is focused to an approximately 0.5 μm spot on the photosensitive layer of the OD to record each pixel of the image sequentially. This method results in the reference images being recorded in binary form. Images with grayscale can be recorded by area modulation of the spots on the OD and read out using the differential interferometric technique.

For example, if binary images of 200×200 pixels are recorded, each image will occupy a 0.2×0.2 mm^2 block. If 0.01 mm spacing is allowed between adjacent blocks, then more than 27,000 reference images can be recorded on a 120-mm-diameter OD. Because the block size is in the order of a fraction of a millimeter, the minute phase variation caused by a nonuniform thickness in plastic cover of the OD can be neglected. To estimate the sequential access time, we assume that the average revolution speed is 1122 rev/min, the average radius is 40 mm, and that all the images are scanned sequentially on each consecutive band of the OD. Then the access time is approximately 40 μs per block. Assuming an optically addressed ferroelectric liquid crystal, SLM is used as the square-law detector, the response time of which is between 10 and 155 μs. Completing one correlation process should take 40–155 μs. This is equivalent as performing more than 6400 correlations per second, a number that can hardly be matched using current electronic technology.

11.3 PHOTOREFRACTIVE-BASED CORRELATOR

Although SLMs can be used to display complex spatial filters, current state-of-the-art SLMs are low-resolution and low-capacity devices. On the other hand, photorefractive (PR) materials offer real-time recording, high resolution, and massive storage capacity. However, thick PR material has limited shift invariant property due to Bragg diffraction [6]. Although thinner crystal can be used, the diffraction efficiency and storage capacity will be substantially reduced. Nevertheless, high-storage capacity, high-diffraction efficiency, and large shift invariance can be achieved by using a reflection-type wavelength-multiplexed PR matched filter, as shown in Fig. 11.7. The matched filter is recorded by combining the Fourier

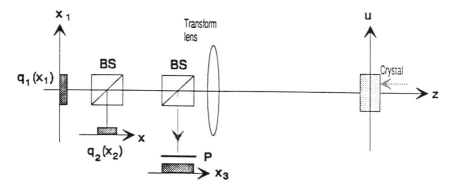

Fig. 11.7 Wavelength-multiplexed reflection-type matched filter correlator. BS's, beam splitters; P, polarizer.

spectrum of an object beam $q_1(x_1)$ with a reference plane wave from the opposite direction. Correlation can be done by inserting an input object $q_2(x_2)$ at plane x, by which the output correlation can be observed at plane x_3. To separate the reading beam $q_2(x_2)$ from the writing beam, the reading beam can be made orthogonally polarized to the writing beam by a polarized beam splitter. By using the coupled mode wave theory, the correlation peak intensity as a function of shift variable S can be shown to be

$$R(S) = \left| q_1^*(x_1)q_1(x_1) \, \text{Sinc}\left[-\frac{\pi D}{n\lambda}\frac{S(x_1 - S)}{f^2}\right] dx_1 \right|^2 \qquad (11.5)$$

where D is the thickness of the crystal. The shift tolerance figure-of-merit (FOM) as

$$(\text{FOM})_{\text{RC}} = XS_{\max} = \frac{4n\lambda}{D}f^2 - S_{\max}^2 \qquad (11.6)$$

where $S_{\max}$ denotes the maximum allowable shift. Plots of FOMs along with the VLC and the JTC are shown in Fig. 11.8, in which we see that the reflection-type PR correlator performs better. It has a higher shift tolerance,

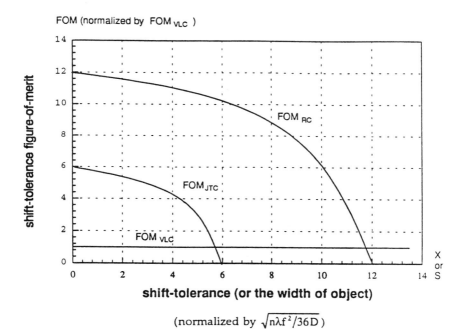

Fig. 11.8 FOMs for different types of correlators.

approximately one order higher than the VLC. The wavelength-multiplexed filter also offers higher and more uniform wavelength selectivity as compared with the angular-multiplexed technique [7].

11.4 OPTICAL NEURAL NETWORKS

Electronic computers can solve computational problems thousands of times faster and more accurately than the human brain. However, for cognitive tasks such as pattern recognition understanding and speaking a language, etc., the human brain is much more efficient. In fact, these tasks are still beyond the reach of modern electronic computers. A neural network (NN) consists of a collection of processing elements, called *neurons*. Each neuron has many input signals, but only one fanned-out signal to many pathways connected to other neurons. These pathways interconnect with other neurons to form a network. The operation of a neuron is determined by a transfer function that defines the neuron's output as a function of the input signals. Every connection entering a neuron has an adaptive coefficient called a *weight* assigned to it. The weight determines the interconnection strength between neurons, and they can be changed through a learning rule that modifies the weights in response to input signals and the transfer function. The learning rule allows the response of the neuron to change with time, depending on the nature of the input signals. This means that the network *adapts* itself to the environment and organizes the information within itself, a type of *learning*.

Generally speaking, a one-layer neural network of N neurons has N^2 interconnections. The transfer function of a neuron can be described by a nonlinear relationship such as a step function, making the output of a neuron either 0 or 1 (binary), or a *sigmoid* function, which gives rise to analog values. The operation of the ith neuron in the network can be represented by a *retrieval equation*, as given by

$$u_i = f\left(\sum_{j=1}^{N} T_{ij}u_j - \theta_i\right) \tag{11.7}$$

where u_i is the activation potential of the ith neuron, T_{ij} is the *interconnection weight* matrix (IWM) (associative memory) between the jth neuron and the ith neuron, and f is a nonlinear processing operator.

Light beams propagating in space will not interfere with each other and optical systems generally have large-space bandwidth products. These are the primary features that prompted the optical implementation of neural networks. The first optical implementation of neural networks was proposed by Psaltis et al. [8]. Since then, a score of optical neural net architectures

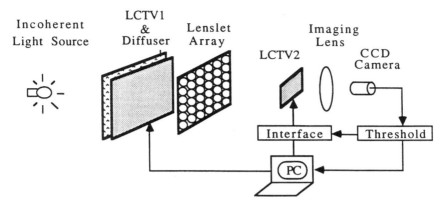

Fig. 11.9 LCTV optical neural network.

have been proposed [9]. A typical hybrid optical neural network using a liquid crystal TV (LCTV) SLM is shown in Fig. 11.9, in which the lenslet array provides the interconnection of the IWM to the input pattern. The transmitted light field after LCTV2 is collected by an imaging lens, focusing at the lenslet array and imaging onto a CCD array detector. The array of detected signals is sent to a thresholding circuit, and the final pattern can be viewed at the monitor or sent back for the next iteration. The data flow is primarily controlled by the microcomputer, for which the LCTV based neural net is indeed an *adaptive* neural network.

One of the most frequently used neural net models is the Hopfield model [10]. The model utilizes an associative memory retrieval process equivalent to an iteratively thresholded matrix-vector outer-product expression as given by

$$
\begin{aligned}
V_i &\to 1 \quad &\text{if } \sum_{j=1}^{N} T_{ij} V_j \geq 0 \\
V_i &\to 0 \quad &\text{if } \sum_{j=1}^{N} T_{ij} V_j < 0
\end{aligned}
\tag{11.8}
$$

where V_i and V_j are binary output and binary input patterns, respectively, and the associative memory operation is written as

$$
T_{ij} = \begin{cases} \sum_{m=1}^{N} (2V_i^m - 1)(2V_j^m - 1) & \text{for } i \neq j \\ 0 & \text{for } i = j \end{cases}
\tag{11.9}
$$

where V_i^m and V_j^m are *i*th and *j*th elements of the *m*th binary vector, respectively.

One of the important aspects of the Hopfield model is the ability to retrieve distorted and partial patterns. For example, a set of English letters was stored in a Hopfield neural net, as shown in Fig. 11.10a. The positive and negative parts of the memory matrix are given in Fig. 11.10b and 11.10c, respectively. If a partial pattern of A is presented to the Hopfield net, a reconstructed pattern of A converges by iteration and is shown in Fig. 11.10d. The construction of the Hopfield neural network is through *intrapattern association*, which ignores the association among the stored exemplars. The alternative is called *interpattern association* (IPA) [11]. By using a simple logic operation, an IPA neural network can be constructed. For example, suppose we have three overlapping patterns as given in the Venn diagram of Fig. 11.11, where common and the special subspaces are defined. If one uses the logic operations

$$
\begin{aligned}
&I = A \wedge \overline{(B \vee C)} \\
&II = B \wedge \overline{(A \vee C)} \quad\quad V = (B \wedge C) \wedge \overline{A} \\
&III = C \wedge \overline{(A \vee B)} \quad\quad VI = (C \wedge A) \wedge \overline{B} \\
&IV = (A \wedge B) \wedge \overline{C} \quad\quad VII = (A \wedge B \wedge C) \wedge \overline{\varphi}
\end{aligned}
\tag{11.10}
$$

then an IPA neural net can be constructed. If the interconnection weights are assigned to 1, −1, and 0, for excitory, inhibitory, and null interconnections, then a tri-state IPA neural net can be constructed. For example, in Fig. 11.12a, pixel 1 is the common pixel among patterns A, B, and C; pixel 2 is the common pixel between A and B; pixel 3 is the common pixel between A and C; whereas pixel 4 is the special pixel, which is also an exclusive pixel with respect to pixel 2. When the logic operations are applied, a tri-state neural network can be constructed as shown in Fig. 11.12b, with the corresponding IPA IWM shown in Fig. 11.11(c).

Pattern translation can be accomplished using the *heteroassociation* IPA neural net [12]. Using similar logic operations among input-output (translation) patterns, a heteroassociative IWM can be constructed. For example, an input-output (translation) training set is given in Fig. 11.13a. Using the logic operations, a heteroassociation neural net can be constructed as shown in Fig. 11.13b, where Fig. 11.13c is its IWM. To illustrate the optical implementation, an input-output training set is shown in Fig. 11.14a. The positive and negative parts of the heteroassociation IWMs are depicted in Fig. 11.14b. If a partial arabic numeral 4 is presented to the optical neural net, a translated Chinese numeral is obtained as shown in Fig. 11.14c. Thus the heteroassociation neural net can indeed translate patterns.

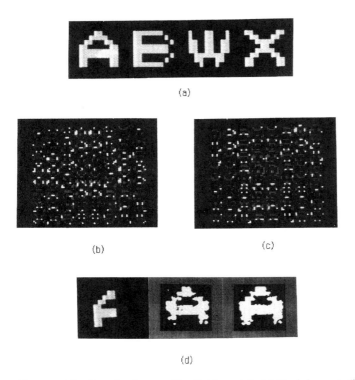

Fig. 11.10 Results from a Hopfield model. (a) Training set; (b) and (c) positive and negative weight matrices, respectively; (d) reconstructed results.

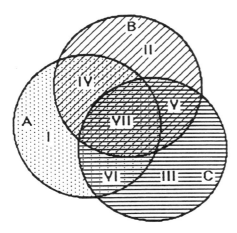

Fig. 11.11 Common and special subspaces.

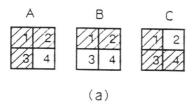

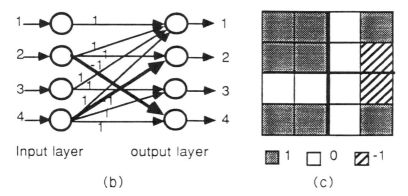

Fig. 11.12 Construction of IPA neural network. (a) Three reference patterns; (b) one-layer neural net; (c) IWM.

11.5 COMPOSITE FILTERS

Classical spatial matched filters are sensitive to rotational and scale variances. A score of approaches to developing composite-distortion-invariant filters [13] have been reported. Among them, the synthetic discriminant function filter (SDF) [14] has played a central role. The idea of SDF can be viewed as a linear combination of classical matched filters, as given by

$$h(x, y) = \sum_{n=1}^{N} a_n f_n(x, y) \tag{11.11}$$

where $\{f_n(x, y)\}$ represents a set of patterns with N different perspective views, and the coefficients $a_n (n = 1, 2, \ldots, N)$ can be determined by solving a set of N simultaneous equations.

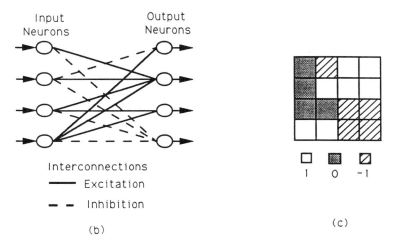

Fig. 11.13 Construction of a heteroassociation IPA neural network. (a) Input-output training sets; (b) a heteroassociation neural net; (c) heteroassociation IWM.

The correlation peak value (CPV) between $h(x, y)$ and an arbitrary training image $f_n(x, y)$ can be shown to be

$$h(x, y) * f_n(x, y)|_{x=0, y=0} = \iint h^*(x, y) f_n(x, y) \, dx \, dy$$
$$= \iint \sum_{n=1}^{N} a_n^* f_n^*(x, y) f_m(x, y) \, dx \, dy \quad (11.12)$$
$$= \sum_{n=1}^{N} a_n^* R_{nm} = C$$

Pattern Recognition with Optics 265

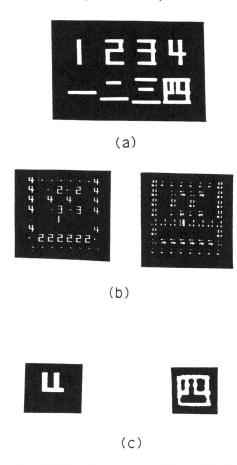

Fig. 11.14 Pattern translation. (a) Arabic-Chinese training set; (b) hetero-association IWM (positive and negative parts); (c) partial arabic numeric to be translated to Chinese numeric.

where C represents a prespecified constant, which is independent of the perspective view of $f_n(x, y)$. In other words, the CPV from an SDF filter is independent of the training images. Figure 11.15 shows a flowchart for an SDF filtering system, where the SDF filter is off-line synthesized while the correlation is on-line processed.

There are techniques proposed to improve the performance of the classical SDF filter [16]. Nevertheless, it is not the intention here to

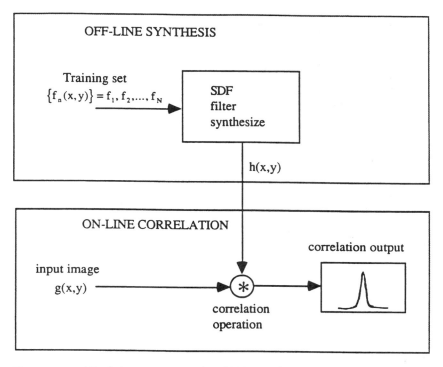

Fig. 11.15 Black-box representation SDF correlator.

describe all those techniques. Instead, two of the simplest approaches will be mentioned: (1) Higher noise tolerance can be achieved if the variance of the SDF filter is minimized, and (2) sharper correlation peaks can be obtained if the average correlation energy of the SDF filter is minimized.

Simulated annealing algorithms (SAA) have also recently been applied to pattern recognition [17, 18]. A spatial domain bipolar filter obtained by SAA [19] can be directly implemented on an input phase-modulating SLM in a JTC. To demonstrate the performance of a bipolar composite filter (BCF), the set of out-of-plane oriented (T72) tanks shown in Fig. 11.16a is used as a training set. The constructed BCF using the SAA is shown in Fig. 11.16b. An input scene to the JTC is shown in Fig. 11.17c, of which the BCF of Fig. 11.16b is used, and the corresponding output correlation distribution as obtained is plotted in Fig. 11.17b. We see that target T72 tanks can indeed be extracted from the noisy background.

Pattern Recognition with Optics 267

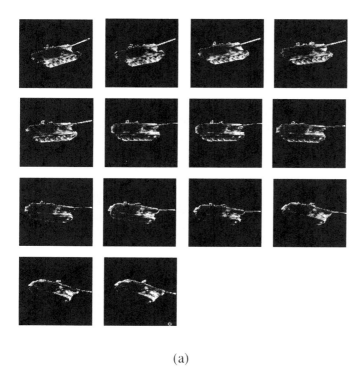

(a)

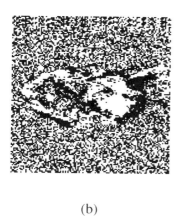

(b)

Fig. 11.16 (a) Out-of-plane rotation T72-tank images; (b) a bipolar composite filter to detect T72 tank.

(a)

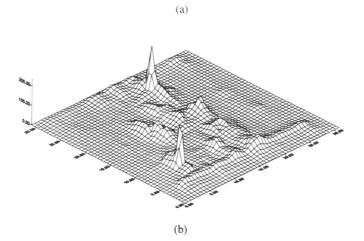

(b)

Fig. 11.17 (a) Input scene; (b) output correlation distribution.

11.6 REMARKS

The synthesis of Fourier domain matched filters using holographic techniques was first initiated by Vander Lugt in 1964 [2]. Since then VLCs have been intensively studied, improved, and modified by advances in electronic-addressable SLMs and nonlinear photorefractive devices. A couple of years ago a different kind of correlator, called the JTC, was demonstrated by Weaver and Goodman [20] and independently by Rau [21]. However, almost no application of the JTC was reported prior to 1984, despite its inclusion in several texts on optical processing. This may have been due to lack of adequate interface devices such as SLMs and CCD cameras. In 1984, Yu and Lu [3] were the first to take advantage of newly available SLMs and demonstrated a programmable JTC. This resulted in the rebirth of the JTC.

Pattern recognition can also be approached using simulated neural networks (NN). The first optical implementation of neural networks was proposed by Psaltis and Farhat in 1985 [22]. Since then neuropattern recognition has seen to a wide variety of optical implementations.

REFERENCES

1. E. Abbe, "Beiträge Zur Theorie des Mikroskops und der Mikroskopischen Wahrnehmung," *Archiv für Mikroskopische Anatomie*, vol. 9, pp. 413–463, 1873.
2. A. Vander Lugt, "Signal Detection by Complex Spatial Filtering," *IEEE Trans. Inf. Th.*, vol. IT-10, pp. 139–145, 1964.
3. F. T. S. Yu and X. J. Lu, "A Real-Time Programmable Joint Transform Correlator," *Opt. Commun.*, vol. 52, pp. 10–16, 1984.
4. A. D. Gara, "Real-Time Tracking of Moving Objects by Optical Correlation," *Appl. Opt.*, vol. 18, pp. 172–174, 1979.
5. F. T. S. Yu, E. C. Tam, T. W. Lu, E. Nishihara, and T. Nishikawa, "Optical Diskbased Joint Transform Correlator," *Appl. Opt.*, vol. 30, pp. 915–916, 1991.
6. F. T. S. Yu, S. Yin, and Z. H. Yang, "Thick Volume Photorefractive Crystal Wavelength-Multiplexed Reflection-Type Matched Filter," *Optical Memory and Neural Networks*, vol. 3, pp. 207–214, 1994.
7. S. Yin, H. Zhou, M. Wen, J. Zhang and F. T. S. Yu, "Wavelength-Multiplexed Holographic Construction Using a Ce:Fe: Doped $LiNbO_3$ Crystal with a Tunable Visible-Light Diode Laser," *Opt. Commun.*, vol. 101, pp. 317–321, 1993.
8. D. Psaltis and N. Farhat, "Optical Information Processing Based on an Associative-Memory Model of Neural Nets with Thresholding and Feedback," *Opt. Lett.*, vol. 19, pp. 98–100, 1985.

9. F. T. S. Yu, T. Lu, X. Yang, and D. A. Gregory, "Optical Neural Network with Pocket-Sized Liquid-Crystal Televisions," *Opt. Lett.*, vol. 15, pp. 863–865, 1989.
10. J. J. Hopfield, "Neurons with Graded Responses Have Collective Computational Properties Like Those of Two-State Neurons," *Proc. Natl. Acad. Sci.*, vol. USA81, pp. 3088–3092, 1984.
11. T. Lu, X. Xu, S. Wu, and F. T. S. Yu, "A Neural Network Model Using Inter-Pattern Association (IPA)," *Appl. Opt.*, vol. 29, pp. 284–288, 1990.
12. F. T. S. Yu, T. Lu, and X. Yang, "Optical Implementation of Hetero-Association Neural Network with Inter-Pattern Association Model," *Int. J. Opt. Comput.*, vol. 1, pp. 129–140, 1990.
13. H. J. Caulfield and R. Haimes, "Generalized Matched Filtering," *Appl. Opt.*, vol. 19, pp. 181–183, 1980.
14. D. Casasent, "Unified Synthetic Discriminant Function Computational Formulation," *Appl. Opt.*, vol. 23, pp. 1620–1627, 1984.
15. B. V. K. Vijaya Kumar, "Minimum Variance Synthetic Discriminant Functions," *J. Opt. Soc. Am.*, vol. A3, pp. 1579–1584, 1986.
16. A. Mahalanobis, B. V. K. Vijaya Kumar, and D. Casasent, "Minimum Average Correlation Energy Filters," *Appl. Opt.*, vol. 26, pp. 3633–3640, 1987.
17. S. Kirpatrick, C. D. Gelatt, and M. P. Vecchi, "Optimization by Simulated Annealing," Science, vol. 220, pp. 671–680, 1983.
18. M. Kim, M. R. Feldman, and C. Guest, "Optimum Encoding of Binary Phase-Only Filters with a Simulated Annealing Algorithm," *Opt. Lett.*, vol. 14, pp. 545–547, 1989.
19. S. Yin, M. Lu, C. Chen, F. T. S. Yu, T. Hudson, and D. McMillen, "Design of a Bipolar Composite Filter Using Simulated Annealing Algorithm," *Opt. Lett.*, vol. 20, pp. 1409–1411, 1995.
20. C. S. Weaver and J. W. Goodman, "A Technique for Optically Convolving Two Functions," *Appl. Opt.*, vol. 5, pp. 1248–1249, 1966.
21. J. E. Rau, "Detection of Differences in Real Distributions," *Opt. J. Soc. Am.*, vol. 56, pp. 1490–1494, 1966.
22. D. Psaltis and N. Farhat, "Optical Information Processing Based on an Associative-Memory Model of Neural Nets with Thresholding and Feedback," *Opt. Lett.*, vol. 10, pp. 98–100, 1985.
23. F. T. S. Yu and S. Jutamulia, *Optical Pattern Recognition*, Cambridge University Press, Cambridge, UK, 1998.

12
Computing with Optics

Reaffirmation of optics parallelism and the development of picosecond and femtosecond optical switches have thrust optics into the nontraditional area of digital computing. The motivation primarily comes from the need for higher-performance general-purpose computers [1]. However, a computer with parallelism requires complicated interconnections, which is difficult to achieve by using wires or microcircuitries. Since both parallelism and interconnection are the inherent trait of optics, it is reasonable to look into the development of a general-purpose optical computer. We shall in this chapter restrict ourselves to a discussion of a few topics of computing with optics.

12.1 LOGIC-BASED COMPUTING

All optical detectors are sensitive to light intensity. Binary numbers 0 and 1 represent dark and bright states. Since we cannot be physically generate 1's (bright) from 0's (dark), there are some difficulties which occur when a logical 1 has to be output from 0's (e.g., NOR, XNOR, NAND). Nevertheless, the shadow-casting method [2] can solve these problems by simply initially encoding 1 and 0 in a dual rail form.

The shadow-casting logic essentially performs all 16 Boolean logic functions based on the combination of the NOT, AND, and OR. For example, 1 and 0 are encoded with four cells, as shown in Fig. 12.1a. The spatially encoded formats A and B are placed in contact, which is equivalent to an AND operation, as shown in Fig. 12.1b. This superposed format is the input-coded image of the optical logic array processor and is set at the input plane. Four spatially distributed light-emitting diodes (LEDs) are employed to illuminate the encoded input. The shadows from each LED will be cast onto an output screen, as shown in Fig. 12.2, for which a decoding mask is needed to extract the true output. The shadow casting

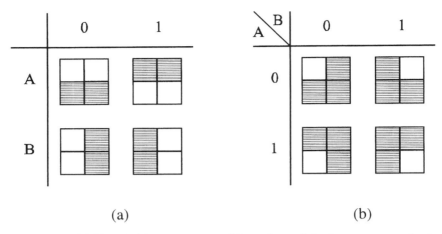

Fig. 12.1 (a) Encoded input patterns; (b) product of the input patterns in the shadow-casting logic array processor.

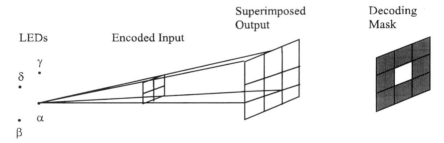

Fig. 12.2 The original shadow-casting logic array processor. The shadows of the encoded input generated by four LEDs are superimposed.

is essentially a selective OR operation among AB, $A\bar{B}$, $\bar{A}B$, and $\overline{AB}$. If the on-off states of the LEDs are denoted by α, β, γ, and δ (where on is 1, and off is 0), the shadow-casting output can be expressed as follows:

$$G = \alpha(AB) + \beta(A\bar{B} + \gamma(\bar{A}B) + \delta(\overline{AB}) \tag{12.1}$$

which is the intensity at the overlapping cells. The complete combination for generating the 16 Boolean functions are tabulated in Table 12.1. A schematic diagram of a hybrid-optical logic array processor is depicted in Fig. 12.3, in which the encoded inputs A and B are displayed on (liquid crystal televisions) LCTV1 and LCTV2, respectively. However, LCTV3

Computing with Optics 273

Table 12.1 Generation of Sixteen Boolean Functions

	α	β	γ	δ
F0	0	0	0	0
F1	1	0	0	0
F2	0	1	0	0
F3	1	1	0	0
F4	0	0	1	0
F5	1	0	1	0
F6	0	1	1	0
F7	1	1	1	0
F8	0	0	0	1
F9	1	0	0	1
F10	0	1	0	1
F11	1	1	0	1
F12	0	0	1	1
F13	1	0	1	1
F14	0	1	1	1
F15	1	1	1	1

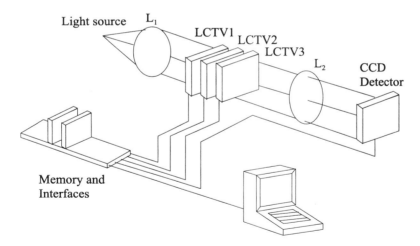

Fig. 12.3 Shadow-casting logic processor using cascaded LCTVs.

(mask) is employed to determine the values of α, β, γ, and δ. The OR operation is performed electronically rather than optically. This has the advantage that no space is needed for shadow-casting NOR. The problems with this system are (1) the encoding operation would slow down the whole

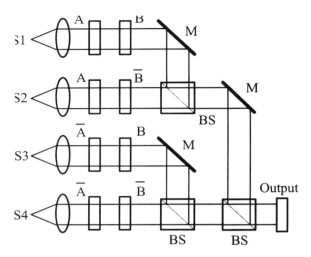

Fig. 12.4 Noncoded shadow-casting logic array processor. S is the light source, BS is the beam splitter, and M indicates a mirror.

process; (2) if the OR operation is performed in parallel, a large number of electronic OR gates and wire intercommunications are required; and (3) if the logic arrays are sequentially performed, a longer processing time is required.

All the shadow-casting logic processors use encoded inputs. If the coding is done by electronics, the overall processing speed will be substantially reduced. On the other hand, the optical output is also an encoded pattern that requires a decoding mask to obtain only the true output. Although the decoding process is parallel and takes no significant processing time, the decoding mask does change the format of the output. A noncoded shadow-casting logic array that is free from these difficulties is shown in Fig. 12.4.

An electronically addressed spatial light modulator (SLM), such as an LCTV, can be used to write an image format. The negation can be done by rotating the LCTV's analyzer (by 90°) without alternating the addressing electronics or software processing. The electronic signal of input A is split into four individual SLMs. Two of them display A, and the other two display $\bar{A}$. Input B follows a similar procedure to generate two B's and two $\bar{B}$'s. The products of AB, $A\bar{B}$, $\bar{A}B$, and $\overline{AB}$ are straightforwardly obtained by putting two corresponding SLMs up to each other. Finally, beam splitters combine the AB, $A\bar{B}$, $\bar{A}B$, and $\overline{AB}$. The logic functions are controlled by the on-off state (α, β, γ, and δ) of the illuminating light (S1, S2, S3, and S4), as illustrated in Table 12.1.

The implementation of these logic functions using LCTVs is straightforward, since no electronic signal modification or software data manipulation is required to encode inputs A and B. Although it seems that more SLMs are needed (four times as many), there is no increase in the space-bandwidth product of the system. In the original shadow-casting logic array, four pixels are used to represent a binary number 1 or 0, while a binary number can be represented by only one pixel in the noncoded system. The use of four sets of individual SLMs is to utilize fully the capability of LCTV to form a negative image format by simply rotating the analyzer by 90°. This method can eliminate the bottleneck of the shadow-casting optical parallel logic array, which is introduced by the coding process.

12.2 OPTICAL-INTERCONNECTS AND SHUFFLING

In addition to logic processing, a digital computer also requires interconnections between processors. It is apparent that when the computer structure becomes more complex, wire interconnections between processors are increasingly difficult to implement. The application of free-space optical interconnection is very likely to replace complicated wire interconnections. In fact, optical interconnect is the most modest, realistic, and near-term application to electronic computers. For example, optical interconnects can be implemented by lenses, fibers, integrated optics, and holograms. The implementations can be used for clock distribution and data communication at different levels, for example, intrachips, interchips, between boards, and among cabinets. For instance, the implementation of perfect shuffle is a typical example of optical interconnect, as will be described below.

Interconnection networks are developed to construct a concurrent processing system by interconnecting a large number of processors and memory modules. The perfect shuffle [3] is an interconnection primitive between local processors for parallel computation. The perfect shuffle is one of the most useful and universal regular interconnects. It is commonly used for sorting, interconnecting array processors, permutation networks, and for special algorithms such as the fast Fourier transform (FFT). An 8-bit perfect shuffle is shown schematically in Fig. 12.5. An optical perfect shuffle can be implemented by diffractive and refractive optics [4], which requires imaging components such as lenses, prisms, gratings, mirrors, and beam splitters, as demonstrated in Fig. 12.6.

For this particular geometry, only two and four windows are required on the first and second mask. The source (left side) and the sink (right side) arrays have the same structure with spacing a. The distance from source

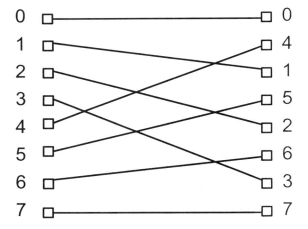

Fig. 12.5 An 8-bit perfect shuffle.

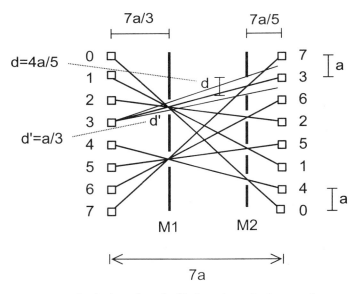

Fig. 12.6 Optical perfect shuffle based on shadow casting.

to sink is $7a$. It is convenient to use a geometrical analysis to determine the position and size of the windows, as well as the location of the two masks. To minimize diffraction effects and to maximize the power transferred, large windows are preferred. However, to keep the system free from cross talk, the

Computing with Optics

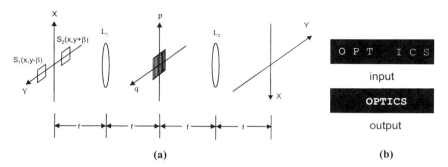

Fig. 12.7 (a) Optical perfect shuffle based on Fourier Optics; (b) input-output result. (*From Ref. 5.*)

width of the window on the second mask must be $d < 4a/5$, and that of the first mask $d' < a/3$. The method can be easily extended to three-dimensional interconnects by using a two-dimensional source and sink arrays. However, the drawback is the power loss due to the fan-out and absorption by the masks.

Another demonstration of a perfect shuffle using an imaging system is depicted in Fig. 12.7a [5]. Two sets of data are designated by $s_1(x, y)$ and $s_2(x, y)$, which are separated by a distance 2β at the input plane. With coherent illumination, the complex light distribution at the input plane with coherent illumination, the complex light distribution at the spatial frequency plane can be described as

$$E(p, q) = S_1(p, q)e^{-\beta q} + S_2(p, q)e^{i\beta q} \tag{12.2}$$

where $S_1(p, q)$ and $S_2(p, q)$ are the Fourier spectra of the input data $s_1(x, y)$ and $s_2(x, y)$ respectively, and (p, q) are the spatial frequency coordinate system. To implement a perfect shuffle, a sinusoidal grating is inserted in the Fourier plane, as given by

$$H(q) = \tfrac{1}{2}\{1 + \cos[(\beta + \Delta\beta)q]\} \tag{12.3}$$

where $\Delta\beta$ is a small displacement constant. Thus, the output light field distribution can be shown as

$$\begin{aligned} I(x, y) = &\tfrac{1}{2}s_1(x, y - \beta) + \tfrac{1}{2}s_2(x, y + \beta) \\ &+ \tfrac{1}{4}[s_1(x, y + \Delta\beta) + s_2(x, y - \Delta\beta)] \\ &+ \tfrac{1}{4}s_1(x, y - 2\beta - \Delta\beta) + \tfrac{1}{4}s_2(x, y + 2\beta + \Delta\beta) \end{aligned} \tag{12.4}$$

in which we see that $S_1(x, y)$ and $S_2(x, y)$ are shifted in opposite directions. If the incremental shift $\Delta\beta$ is equal to the separation between the processing data, S_1 and S_2, then it would be perfectly interlaced, as shown in Fig. 12.7b.

12.3 MATRIX-VECTOR MULTIPLICATION

The optical matrix-vector multiplier can be implemented as shown in Fig. 12.8. The elements of the pattern vector are entered in parallel by controlling the intensities of N light-emitting diodes (LEDs). Spherical and cylindrical lenses L_1 and L_2 combine to image the LED array horizontally onto the matrix mask M, which consists of $N \times N$ elements. The combination of cylindrical and spherical lenses L_3 and L_4 collects all the light from a given row and brings it into focus on one element of the photodetectors. Since each photodetector measures the value of one element of the output vector, we see that it essentially performed the matrix-vector multiplication. This configuration can be further simplified by fabricating line-shaped LEDs and line-shape detector arrays, as depicted in Fig. 12.9. Note that the LED array can be replaced by an SLM with uniform illumination.

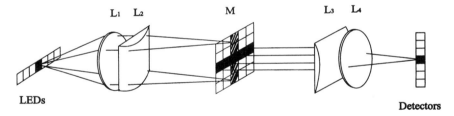

Fig. 12.8 Schematic diagram of an optical matrix-vector multiplier.

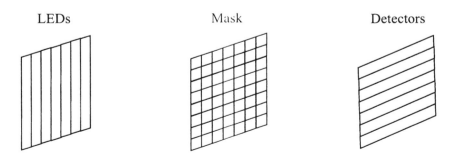

Fig. 12.9 Optical matrix-vector multiplier using lineshape LEDs and detectors.

Important problems can be solved with iterative multiplication of vector by matrix, which includes finding eigenvalues and eigenvectors, solving simultaneous linear equations, computing the discrete Fourier transform, implementation of neural networks, etc. For example, a neural network consists of several layers of neurons, in which two successive neuron layers are connected by an interconnection net. If the neuron structure in a layer is represented by a vector, then the interconnection can be represented by a matrix. For example, Hopfield neural network [6] uses an associative memory retrieval process, which is essentially a matrix-vector multiplier as given by

$$V_i \to 1 \quad \text{if } \Sigma T_{ij} V_j \geq 0$$
$$V_i \to 0 \quad \text{if } \Sigma T_{ij} V_j < 0 \tag{12.5}$$

where V_i and V_j are the output and the input binary vectors, respectively, and T_{ij} is the interconnection matrix.

We note that for binary matrix, the matrix-vector multiplier is in fact a *crossbar switch*. The crossbar switch is a general switching device that can connect any N input to any N output; this is called *global connection*. Crossbar switches are usually not implemented in electronic computers because they would require N^2 individual switches, but they are used in telephone exchanges. On the other hand, an optical crossbar interconnected signal processor would be very useful for performing fast Fourier transforms (FFT), convolution and correlation operations by taking advantage of the reconfigurability and parallel processing of crossbar interconnection [7]. Also, the optical crossbar switch can be employed to implement a programmable logic array (PLA) [8]. The electronic PLA contains a two-level, AND-OR circuit on a single chip. The number of AND and OR gates and their inputs is fixed for a given PLA. A PLA can be used as a read-only memory (ROM) for the implementation of combinational logic.

12.4 SYSTOLIC PROCESSOR

The engagement matrix-vector multiplier is in fact a variation of a systolic processor, as illustrated in Fig. 12.10 [9]. The components of vector **B** are shifted into the multiplier-adder modules starting at time t_0. Subsequent vector components are clocked in contiguously at t_1 for b_2, t_2 for b_3, and so on. At time t_0, b_1 is multiplied with a_{11} in module 1. The resultant $b_1 a_{11}$ is retained within the module to be added to the next product. At time t_1, b_1 is shifted to module 2 to multiply with a_{21}. At the same time, b_2 enters module 1 to multiply with a_{12}, which forms the second product of the output

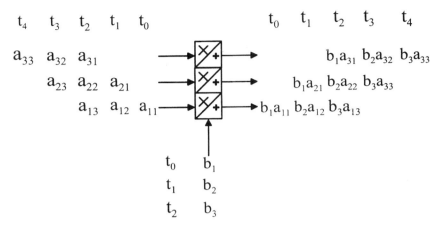

Fig. 12.10 Conceptual diagram of engagement systolic array processor. (*From Ref. 9.*)

vector component. Consequently, module 1 now contains the sum $b_1 a_{11} + b_2 a_{12}$. This process continues until all the output vector components have been formed. In all, $2N - 1$ clock cycles that employ N multiplier-adder modules are required. The main advantage of a systolic processor is that optical matrix-vector multiplication can be performed in the high-accuracy digital mode [10].

A discrete linear transformation (DLT) system can be characterized by an impulse response h_{mn}. The input-output relationship of such a system can be summarized by

$$g_m = \sum f_n h_{mn} \qquad (12.6)$$

Since the output g_m and the input f_n can be considered vectors, the preceding equation can be represented by a matrix-vector multiplication, where h_{mn} is known as a *transform matrix*. Thus, the different DLTs would have different matrices. The discrete Fourier transform (DFT) is one of the typical examples for DLT, as given by

$$F_m = \frac{1}{N} \sum f_n \exp \frac{-i2\pi mn}{N} \qquad (12.7)$$

where $n = 0, 1, \ldots, N$, and

$$h_{mn} = \exp \frac{-i2\pi mn}{N} \qquad (12.8)$$

Computing with Optics

is also known as the *transform kernel*. To implement the DFT transformation is an optical processor, we present the complex transform matrix with real values, for which the real transform matrices can be written as

$$\text{Re}[h_{mn}] = \cos\frac{2\pi mn}{N} \quad I_m[h_{mn}] = \sin\frac{2\pi mn}{N} \tag{12.9}$$

which are the well-known *discrete consine transform* (DCT) and *discrete sine transform* (DST).

The relationship between the real and imaginary parts of an analytic signal can be described by the *Hilbert transform*. The discrete Hilbert transform (DHT) matrix can be written as

$$\begin{aligned} h_{mn} &= 2\frac{\sin^2[\pi(m-n)/2]}{\pi(m-n)} \quad m-n \neq 0 \\ &= 0 \quad m-n = 0 \end{aligned} \tag{12.10}$$

Another frequently used linear transform is the Chirp-Z transform, which can be used to compute the DFT coefficients. The discrete Chirp-Z transform (DCZT) matrix can be written as

$$h_{mn} = \exp\frac{i\pi(m-n)^2}{N} \tag{12.11}$$

Since the DLT can be viewed as the result of a digital matrix-vector multiplication, systolic processing can be used for the implementation. By combining the systolic array processing technique and the *two-complement representation*, a DLT can be performed with a digital optical processor. As compared with the analog optical processor, the technique offers high accuracy with low error rate. Also, it is compatible with other digital processors.

Two's complement representation can be applied to improving the accuracy of matrix multiplication. Two's complement numbers provide a binary representation of both positive and negative values, and facilitate subtraction by the same logic hardware that is used in addition. In a b-bit two's complement number, the most significant bit is the sign. The remaining $b-1$ bits represent its magnitude. An example for DCT is given in Fig. 12.11, in which the digital transform matrix has been encoded in two's complement representation with systolic engagement format.

Discrete Cosine Transform (DCT)

$$[h_{m,n}] = \begin{pmatrix} \cos[(2\pi/3)*0*0] & \cos[(2\pi/3)*0*1] & \cos[(2\pi/3)*0*2] \\ \cos[(2\pi/3)*1*0] & \cos[(2\pi/3)*1*1] & \cos[(2\pi/3)*1*2] \\ \cos[(2\pi/3)*2*0] & \cos[(2\pi/3)*2*1] & \cos[(2\pi/3)*2*2] \end{pmatrix}$$

$$= \begin{pmatrix} 1 & 1 & 1 \\ 1 & -1/2 & -1/2 \\ 1 & -1/2 & -1/2 \end{pmatrix} = \begin{pmatrix} 00100 & 00100 & 00100 \\ 00100 & 11110 & 00010 \\ 00100 & 11110 & 11110 \end{pmatrix}$$

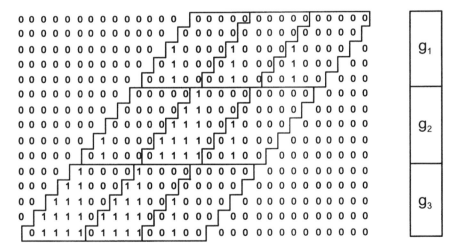

Fig. 12.11 Transformation matrix for discrete cosine transform using two's complement representation.

12.5 MATRIX-MATRIX PROCESSING

The matrix-matrix multiplier is a mathematical extension from the matrix-vector multiplier. In contrast, the implementation of the matrix-matrix multiplier requires a more complex optical setup. Matrix-matrix multipliers may be needed to change or process matrices that

Computing with Optics

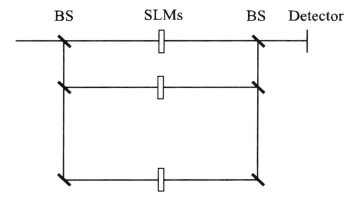

Fig. 12.12 Matrix-matrix multiplier based on outer product.

will eventually be used in a matrix-vector multiplier. Matrix-matrix multiplication can be computed by successive outer-product operations as follows:

$$\begin{bmatrix} a_{11} & a_{12} & a_{13} \\ a_{21} & a_{22} & a_{23} \\ a_{31} & a_{32} & a_{32} \end{bmatrix} \begin{bmatrix} b_{11} & b_{12} & b_{13} \\ b_{21} & b_{22} & b_{23} \\ b_{31} & b_{32} & b_{33} \end{bmatrix} = \begin{bmatrix} a_{11} \\ a_{21} \\ a_{31} \end{bmatrix} [b_{11} \quad b_{12} \quad b_{13}]$$

$$+ \begin{bmatrix} a_{12} \\ a_{22} \\ a_{32} \end{bmatrix} [b_{21} \quad b_{22} \quad b_{23}] + \begin{bmatrix} a_{13} \\ a_{23} \\ a_{33} \end{bmatrix} [b_{31} \quad b_{32} \quad b_{33}]$$

(12.12)

Since the outer-product an beexpressed as

$$\begin{bmatrix} a_{11} \\ a_{21} \\ a_{31} \end{bmatrix} [b_{11} \quad b_{12} \quad b_{13}] = \begin{bmatrix} a_{11}b_{11} & a_{11}b_{12} & a_{11}b_{13} \\ a_{21}b_{11} & a_{21}b_{12} & a_{21}b_{13} \\ a_{31}b_{11} & a_{31}b_{12} & a_{31}b_{13} \end{bmatrix} \qquad (12.13)$$

it can be obtained by simply butting two SLMs. Equation (12.13) can be realized optically as depicted in Fig. 12.12, in which each pair of SLMs performs the multiplication of the outer products, while the beam splitters perform the addition of the outer products. Thus we see that the whole combinational operation can be completed in one cycle. However, if the addition is performed electronically in a sequential operation, then only a pair of SLMs is required. If the multiplication is performed sequentially,

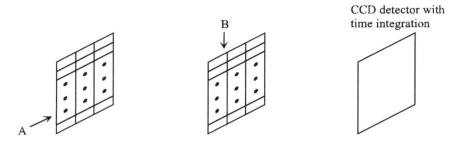

Fig. 12.13 Matrix-matrix multiplier based on systolic array.

then more than one cycle operation is needed. This approach is known as systolic processing. The obvious advantage of systolic processing is that fewer SLMs are needed. However the tradeoff is the increase in processing time.

Figure 12.13 illustrates an example of systolic array matrix operation. Two transmission-type SLMs A and B are placed close together at the input plane. By successively shifting the A and B systolic array matrix formats, one can obtain the product of matrices A and B using a time-integrating CCD detector at the output plane. Although only two SLMs are required, the computation time needed for performing an $n \times n$ matrix-matrix multiplication of b-bit number would be $(2nb-1)+(n-1)$ times that needed by using an outer-product processor.

Figure 12.14 shows an example of a systolic outer-product processing. The optical processor consists of $n \times n = 3 \times 3 = 9$ pieces of SLM with $b \times b = 5 \times 5$ pixels each. Note that the systolic array representation of matrices A and B differs from the systolic formats described previously. By sequentially shifting the row and column elements of A and B into the SLMs, we can implement the $a_{ij}b_{kl}$ multiplication at each step, with an outer-product operation that has been performed in the 5×5 SLM. The result can be integrated in relation to time with a CCD detector at the output plane. Since more SLMs are employed in parallel, the matrix-matrix multiplication can be completed in fewer steps.

12.6 EXPERT SYSTEM AND ARTIFICIAL INTELLIGENCE

An expert system is one of the most common applications of artificial intelligence (AI). Expert systems have been developed to assist complex planning, scheduling tasks, diagnosing diseases, locating mineral deposits,

Computing with Optics 285

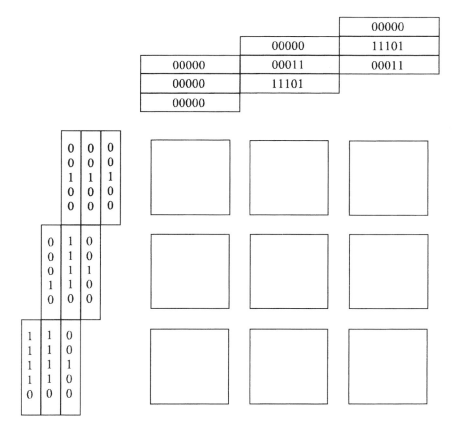

Fig. 12.14 Matrix-matrix multiplication based on the systolic outer-product method.

and so forth [11]. A popular approach to an expert system is to use IF-THEN rules. These rules say that if a certain kind of situation arises, then a certain kind of action can be taken. There are two basic control strategies for applying the IF-THEN rules. The first strategy is to select a goal to be achieved and then to scan the rules backward to find those consequences that can achieve the goal. This is known as *goal-driven backward chaining*. Most existing expert systems use such a backward-chaining strategy. A different strategy is to scan forward through the rules until one is found whose antecedents match assertions in the data base. The rule is applied, updating the data base, and the scanning resumes. This process continues until either a goal state is reached or no applicable rules are found. This strategy is

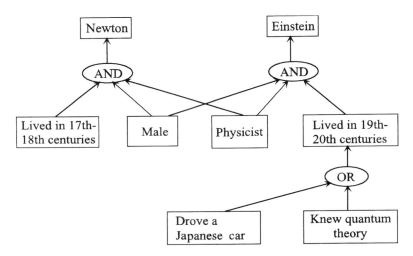

Fig. 12.15 Example of AND/OR graph for an expert system.

known as *data-driven forward chaining*. The data-driven forward-chaining strategy may approach the ideal parallel system, and would be suitable for an optical inference engine. In other words, the *sequential* reasoning electronic-inference engine is replaced by a *parallel* reasoning optical-inference engine. A relatively simple architecture for an optical-inference engine based on parallel propositional logic is described as follows [12]. The problem to be solved by an expert system is usually represented by an AND/OR graph, which is a network combining a number of IF-THEN rules. Basically, only AND and OR gates are required to performed the inferences. A simple example of an AND/OR graph is shown in Fig. 12.15. LCTVs can be used to implement the parallel data-driven forward-chaining strategy for the AND/OR graph shown in the figure. Needless to say, a hybrid optical expert system utilizing LCTVs to perform inference processing can be employed.

Let is consider that off-pixel and on-pixel LCTVs have 1 and 2 transmittances in an arbitrary unit. The threshold level for the resultant transmittance can be set to determine the logic function as either AND or OR. For example, if the threshold is set at 2, and OR function is obtained, and the AND function can be obtained by setting the threshold equal to 3. The implementation of AND/OR graph is illustrated in Fig. 12.15, in which three LCTVs are arranged in tandem to form three-input AND or OR gates. The resultant transmittances are listed in Table 12.2, for which we see that the threshold selections at 2 and 6 result in OR and AND

Computing with Optics

Table 12.2 Resultant Transmittance of Three LCTVs

LCTV1	LCTV2	LCTV3	Resultant	OR	AND
1 (off)	1 (off)	1 (off)	1	off	off
1 (off)	1 (off)	2 (on)	2	on	off
1 (off)	2 (on)	1 (off)	2	on	off
1 (off)	2 (on)	2 (on)	4	on	off
2 (on)	1 (off)	1 (off)	2	on	off
2 (on)	1 (off)	2 (on)	4	on	off
2 (on)	2 (on)	1 (off)	4	on	off
2 (on)	2 (on)	2 (on)	8	on	on

functions, respectively. Notice it is a hybrid system, a computer is used to store, update the knowledge base, to write symbols (logic patterns) on the LCTVs, and to control the whole operation. The search based on the AND/OR graph shown in Fig. 12.15 may be completed in two cycles. Each cycle performs one to two inference rules in parallel. The knowledge based is updated after each cycle. An LCTV display consists of about 140 × 120 pixels. To perform less than 140 inference rules in parallel, a vertical line can be used instead of a pixel to represent a fact. A cylindrical lens can be used to focus the vertical line onto an element of the array detector. This optical setup further introduces fault tolerance due to defects associated with the LCTV and noise. The detected signal is fed back into a computer to be thresholded and to complete one cycle. Let us detail the operation of the preceding expert system in the following.

For example, the given facts are: DROVE A JAPANESE CAR; KNEW QUANTUM THEORY; PHYSICIST; MALE; LIVED IN 17TH–18TH CENTURIES; as the inputs to the computer to generate a symbolic representation. The intermediate results from a cycle may be used to update the knowledge base that is needed for the subsequent inference cycles. For instance, results from the first cycle are used to generate symbols for the second cycle. For a given example—DROVE A JAPANESE CAR = NO; KNEW QUANTUM THEORY = YES; MALE = YES; LIVED IN 17TH–18TH CENTURIES = NO—the generated symbols are illustrated in Fig. 12.16. If a logic gate is operated by only two inputs, then the other input is set off, and the thresholds for the OR and AND functions are 2 and 3, respectively. The first cycle will update the knowledge base with the fact LIVED IN 19TH–20TH CENTURIES = YES, which shows that the fact EINSTEIN is finally YES, while the other alternative is NO.

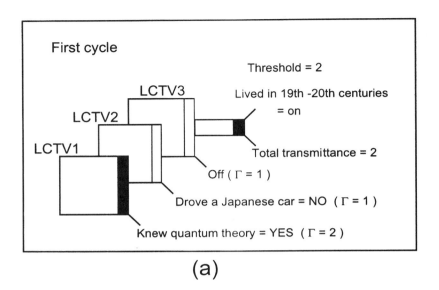

(a)

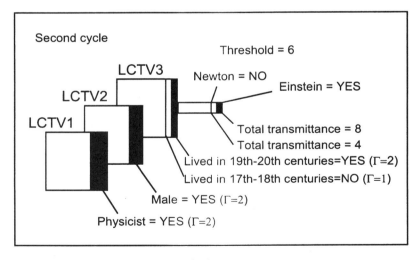

(b)

Fig. 12.16 Operation of inference. Symbols are generated on LCTVs at (a) the first cycle, and (b) the second cycle. Black stripes represent on-pixels (transmittance = 2); white stripes represent off-pixels (transmittance = 1).

In the aforementioned example, when the fact MALE is YES, the chances for NEWTON and EINSTEIN are the same. In contrast, medical diagnostic expert systems are usually based on the probabilistic principle. If a symptom is positive, there are uneven probabilities for different diseases. The optical expert system [13] for diagnosis is used, the database is to establish a list of *a priori* probabilities $p(e_i/h_j)$, that an event of symptom e_i will be present given a specified hypothesis or illness h_j. The expert system is then used to compute the a posteriori probabilities $p(h_j/e_i)$, of a specified illness h_j, given a specified symptom e_i. After the patient has answered a number of questions, the illness or illnesses with the highest probabilities are the most likely hypotheses as described by the following *Bayes' rule*:

$$p(h_j/e_i) = \frac{p(e_i/h_j)p(h_j)}{p(e_i/h_j)p(h_j) + p(e_i/\bar{h}_j)p(\bar{h}_j)} \tag{12.14}$$

where $\bar{h}_j$ is not $-h_j$, and $p(\bar{h}_j) = 1 - p(h_j)$. The rule can be extended to multiple events and circumstances. Two terms in the denominator (one is identical to the numerator) can be obtained using optical matrix-vector multipliers. If a response to the ith event has been received, a unit signal is entered at this input, while all other inputs remain zero. Notice that these inputs form the input vector, while the a prior probabilities $p(e_i/h_i)$ form the interconnection matrix. The elements of the output vector are multiplied by the corresponding $p(h_i)$, which are known from statistics of the region and the patient profile.

Both the deterministic and probabilistic examples given in the preceding are based on relatively simple proposition calculus, in which a true or false fact is simply a statement without any arguments. Optical expert systems based on predicate calculus have been proposed [14, 15]. These systems are based on the production system paradigm, where facts and rules constitute a knowledge base. Facts define the relationship between objects, and rules define the procedures for deriving new relations from existing ones. PROLOG (which has been selected as the basis for the Japanese fifth-generation computers) is a popular AI language that implements the production system paradigm.

AI systems generally consist of a knowledge base and an inference engine. The inference engine accepts as input a set of facts and a set of rules from the knowledge base and one or more queries. The output of the inference engine is a set of specific conclusions that are logically inferred from the facts and rules in response to the queries. For example, a set of data could be a set of names of people.

$$D = \{\text{Ann, Beth, Carol, David, Edward, Frank}\} \tag{12.15}$$

A set of possible relationships for D among them is given

$$R = \{\text{married-to, mother-of, father-of, son-of, daughter-of, child-of,}$$
$$\text{is-male, is-female}$$

(12.16)

The data objects and relationships link among them is a collection of facts and rules that relate the elements of D and R, for instance,

David married-to Ann
Edward married-to Beth
David father-of Carol
Edward father-of Frank

The query addressed to this knowledge base is "Who is the mother of Carol?" or ?mother-of Carol. The PROLOG code for these expressions is given as follows:

married-to (David, Ann)
married-to (Edward, Beth)
father-of (David, Carol)
father-of (Edward, Frank)

and the query

?-mother-of (X, Carol)

In an optical-inference engine [14], mapping templates are used to store the relationships between the data objects, and are thus defined by the facts. Conclusions are assigned to queries by applying these mapping templates to the data objects in the order prescribed by the rules. The married-to and father-of facts can be represented by two-dimensional mapping templates, as shown in Fig. 12.17a and b, respectively. Operation of the ?mother-of Carol query is performed as follows.

An LCTV is used to display the mapping template for the father-of facts, that is, Fig. 12.17a. The query data Carol will turn on a horizontal line of C to illuminate the LCTV which is in fact a two-dimensional optical AND operation between the horizontal line. The transmittance from the LCTV will be low, except that the element (D, C) being high indicates that David is the father of Carol. The bright spot at (D, C) is then expanded optically (using a cylindrical lens) to form a vertical line, and is projected onto another LCTV showing the mapping template of the married-to facts, i.e., Fig. 12.17b. Consequently, the transmittance from the second LCTV shows only one bright spot at (D, A). The bright spot is now expanded

Computing with Optics

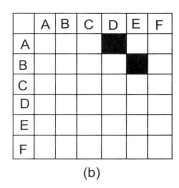

Fig. 12.17 Templates showing the relationship of (a) father-of and (b) married-to.

horizontally so that it can be detected by the detector A. This indicates the conclusion is Ann.

It is interesting to notice that the mapping template method can also be performed using an algebraic matrix operation [15] for which the married-to facts can be represented by the following binary matrix:

$$[MT] = \begin{array}{c} \\ A \\ B \\ C \\ D \\ E \\ F \end{array} \begin{array}{c} A\ B\ C\ D\ E\ F \\ \begin{bmatrix} 0 & 0 & 0 & 0 & 0 & 0 \\ 0 & 0 & 0 & 0 & 0 & 0 \\ 0 & 0 & 0 & 0 & 0 & 0 \\ 1 & 0 & 0 & 0 & 0 & 0 \\ 0 & 1 & 0 & 0 & 0 & 0 \\ 0 & 0 & 0 & 0 & 0 & 0 \end{bmatrix} \end{array}$$

and the father-of fact is represented by another binary matrix:

$$[FO] = \begin{array}{c} \\ A \\ B \\ C \\ D \\ E \\ F \end{array} \begin{array}{c} A\ B\ C\ D\ E\ F \\ \begin{bmatrix} 0 & 0 & 0 & 0 & 0 & 0 \\ 0 & 0 & 0 & 0 & 0 & 0 \\ 0 & 0 & 0 & 0 & 0 & 0 \\ 0 & 0 & 1 & 0 & 0 & 0 \\ 0 & 0 & 0 & 0 & 0 & 1 \\ 0 & 0 & 0 & 0 & 0 & 0 \end{bmatrix} \end{array}$$

The query ?mother-of Carol can be answered by the rule:

X mother-of Carol, IF Y married-to X AND Y father-of Carol

(12.17)

which involves two facts connected by a rule. One may consider that, in general, a rule is a function of facts. A new fact can be concluded from the function, through a sequence of logic operations applied to the known facts. In the matrix-encoding method, logic operations between relations are carried out by thresholding the matrix-matrix product for the AND operation and by thresholding the matrix-matrix sum for the OR operation. Equation (12.17) can be used to generate new facts of mother-of as follows:

$$[MO] = [MT][FO] \tag{12.18}$$

which yields

$$[MO] = \begin{array}{c} \\ A \\ B \\ C \\ D \\ E \\ F \end{array} \begin{array}{c} A & B & C & D & E & F \\ \begin{bmatrix} 0 & 0 & 1 & 0 & 0 & 0 \\ 0 & 0 & 0 & 0 & 0 & 1 \\ 0 & 0 & 0 & 0 & 0 & 0 \\ 0 & 0 & 0 & 0 & 0 & 0 \\ 0 & 0 & 0 & 0 & 0 & 0 \\ 0 & 0 & 0 & 0 & 0 & 0 \end{bmatrix} \end{array}$$

To get the conclusion of the query ?mother-to Carol, a matrix-vector multiplication is performed, in which the input vector is $[A=0, B=0, C=1, D=0, E=0, F=0.]$ The output vector is obtained as follows:

$$\begin{bmatrix} 0 & 0 & 1 & 0 & 0 & 0 \\ 0 & 0 & 0 & 0 & 0 & 1 \\ 0 & 0 & 0 & 0 & 0 & 0 \\ 0 & 0 & 0 & 0 & 0 & 0 \\ 0 & 0 & 0 & 0 & 0 & 0 \\ 0 & 0 & 0 & 0 & 0 & 0 \end{bmatrix} \begin{bmatrix} 0 \\ 0 \\ 1 \\ 0 \\ 0 \\ 0 \end{bmatrix} = \begin{array}{c} A \\ B \\ C \\ D \\ E \\ F \end{array} \begin{bmatrix} 1 \\ 0 \\ 0 \\ 0 \\ 0 \\ 0 \end{bmatrix} \tag{12.19}$$

Since the output vector is $[A=1, B=0, C=0, D=0, E=0, F=0]$, the conclusion is Ann. Thus we see that the optical matrix-vector and matrix-matrix multipliers can also be applied to artificial intelligence.

12.7 REMARKS

We have shown that computing can be carried out with optics; however, the proposed optical systems are basically emulating the electronic counterpart for computing. Unfortunately, the available electrooptic devices are still way below the required specifications for this application. The success of optical computing is very much dependant on the future development of these devices. However, such devices will take years to develop or may never develop. Unless the issues of devices can be solved, it hardly can be claimed that we actually have an optical computer to compete.

Nevertheless, there is an alternative approach that remains to be seen. Instead of emulating the electronics for computing, one can search for a new type of mathematics, or develop one, that is more suitable for optics to compute. It is my belief that the behavior of optics and electronics is quite different. For instance, light is a better carrier for information transmission, while electrons interact more efficiently. To ask optics to behave like electronics for computing may not be the most efficient approach.

REFERENCES

1. M. J. Quinn, *Designing Efficient Algorithms for Parallel Computers*, McGraw-Hill, New York, 1987.
2. J. Tanida and Y. Ichioka, "Optical Logic Array Processor Using Shadowgrams," *J. Opt. Soc. Am.*, vol. 73, p. 800, 1983.
3. C. L. Wu and T. Y. Feng, "The Universality of the Shuffle-Exchange Network," *IEEE Trans. Comput.*, vol. C-30, p. 324, 1981.
4. A. W. Lohmann, W. Stork, and G. Stucke, "Optical Perfect Shuffle," *Appl. Opt.*, vol. 25, p. 1530, 1986.
5. Q. W. Song and F. T. S. Yu, "Generalized Perfect Shuffle Using Optical Spatial Filtering," *Appl. Opt.*, vol. 27, p. 1222, 1988.
6. J. J. Hopfield, "Neural Networks and Physical Systems with emergent Collective Computational Abilities," *Proc. Natl. Acad. Sci. U.S.A.*, vol. 79, p. 2554, 1982.
7. A. D. McAulay, "Optical Crossbar Interconnected Digital Signal Processor with Basic Algorithms," *Opt. Eng.*, vol. 25, p. 82, 1986.
8. R. Arrathoon, "Historical Perspectives: Optical Crossbars and Optical Computing," *Proc. SPIE*, vol. 752, p. 2, 1987.
9. H. T. Kung, "Why Systolic Architecture?" Computer, vol. 15, p. 37, 1982.
10. H. J. Caulfield, W. T. Rhodes, M. J. Foster, and S. Horvitz, "Optical Implementation of Systolic Array Processing," *Opt. Commun.*, vol. 40, p. 86, 1981.
11. P. S. Guilfoyle, "Systolic Acousto-Optic Binary Convolver," *Opt. Eng.*, vol. 23, p. 20, 1984.
12. F. Hayes-Roth, D. A. Waterman, and D. B. Lenat, eds. *Building Expert Systems*, Addison-Wesley, Reading, Mass., 1983.
13. S. Jutamulia, G. Storti, X. Li, "Expert Systems Based On LCTV AND/OR Logic," *Opt. Laser Technol.*, vol. 21, p. 392, 1989.
14. A. D. McAulay, "Real-Time Optical Expert Systems," *Appl. Opt.*, vol. 25/26, p. 1927, 1987.
15. C. Warde and J. Kottas, "Hybrid Optical Inference Machines: Architectural Considerations," *Appl. Opt.*, vol. 25, p. 946, 1986.

16. J. Y. Jau, F. Kiamilev, Y. Fainman, S. Esener, and S. H. Lee, "Optical Expert System Based On Matrix-Algebraic Formulation," *Appl. Opt.*, vol. 27, p. 5170, 1988.
17. F. T. S. Yu and S. Jutamulia, *Optical Signal Processing, Computing, and Neural Networks*, Wiley-Interscience, New York, 1992.

13
Communication with Fiber Optics

Exploitation of information transmission by light is one of the oldest and most important aspects of communication for all the living species in this planet earth. In fact, information transfer from one place to another can be easily carried by light from very close range such as microscopic observation to millions of miles by telescopic lenses. Due to the discovery of high-coherence light sources of lasers and the advent of high-quality fiber-optic technologies, communication with optics can be carried out beyond the transoceanic distances with small-size optical fiber. The carrier frequency of fiber optic communications can be as high as hundreds of Thz in visible or in near-infrared range of electromagnetic spectrum. In fact, fiber-optic communication carrier frequency is about 5 orders higher than the conventional microwave communication technology, which is about 1 Ghz region. The objective of this chapter is to describe briefly the basic aspects of communication with fiber optics.

13.1 ASPECTS OF FIBER-OPTIC COMMUNICATION

Basically, the capacity of communication is limited by the bit rate-distance product BL, where B is the bit rate and L is the distance of communication before the signal diminishes to an unrealistic level. In other words, L is the distance between the repeaters. In order to appreciate the significance of using fiber-optic technology for communication, BL product (which represents the transmission capacity) for various emerging telecommunication technologies is plotted as a function of calendar year, as shown in Fig. 13.1. We see that the light-wave technology offers a much higher transmission capacity than the microwave communication. Added with the recent development of the Erbium-doped fiber amplifier, the capacity can be as high as $10^{12} - 10^{15}$ (bits/s)-km.

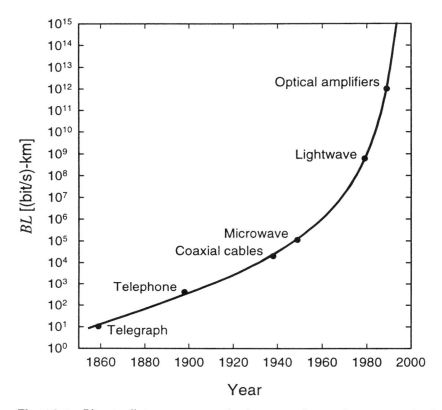

Fig. 13.1 Bit-rate-distance versus calendar-year of emerging communication technologies.

The root for using the light-wave technology for long-distance transmission could have started from the beginning of the discovery of the laser in the 1960s. The breakthrough research on low transmission-loss fiber materials in the 1970s also provided a big surge application on long-distance communication. The transmission loss of fiber-optic can be reduced to about 20 dB/km near the 1 μm wavelength region. In addition, by the recent development of compact GaAs semiconductor lasers, fiber-optic communication has a worldwide commitment to develop highly efficient long-distance communication systems. Although the implementations of fiber-optic communication was postulated [1] as early as the 1970s, it was the actual experimental demonstration in 1980s that showed that transmission can be implemented over 4000 km, by compensating the fiber loss through stimulated Raman scattering [2]. Along with the recent development of Erbium-doped fiber amplifiers, fiber-optic communication systems can in principle transmit information over 1000 km at 10 Gb/s and over 350 km at 20 Gb/s [3]. Mention should also be made that by using recirculating-loop configuration, solitons at 2.4 Gb/s can be maintained over 12,000/km [4]. Although communication technology with fiber optics is barely $2\frac{1}{2}$ decades old, it has reached to the stage of worldwide practical implementations.

In a practical communication, information is generally transmitted with *analog* form, for example, speeches and images. These informational signals can in turn be carried by coded *digital* signals, to improve the signal-to-noise ratio transmission. An example of analog-to-digital signal transmission is illustrated in Fig. 13.2, in which the signal sampling is determined by the *Nyquist* rate. The next step is to encode the sampled value into binary digital form, for which each sampled value is quantized into "1" and "0" binary formats. The number of m bits sampled value is related to the quantized signal levels m as given by

$$M = 2^m \quad \text{or} \quad m = \log_2 M \tag{13.1}$$

The bit rate associated with the pulse-code-modulation (PCM) would be

$$B = mf_s \geq (2\Delta f)\log_2 M \tag{13.2}$$

where f_s is the Shannon's sampling frequency:

$$f_s \geq \Delta f \tag{13.3}$$

and Δf is the bandwidth of the analog signal. Since this quantization also adds noises, known as *quantization noise*, to this coded signal, the number of quantized levels should be limited by

$$M \geq \frac{A_{\max}}{A_N} \tag{13.4}$$

where $A_{\max}$ is the highest amplitude of the analog signal and A_N is the *root-mean-square* (rms) noise amplitude. Notice that the ratio of $A_{\max}/A_N$ is known as the *dynamic range* of the quantization and is related to the signal-to-noise (SNR) given by

$$SNR = 20\log_{10}\frac{A_{\max}}{A_N} \tag{13.5}$$

In view of the preceding PCM bit rate, we see that

$$B > \frac{\Delta f(SNR)}{3} \tag{13.6}$$

where SNR is in decibel (dB) units.

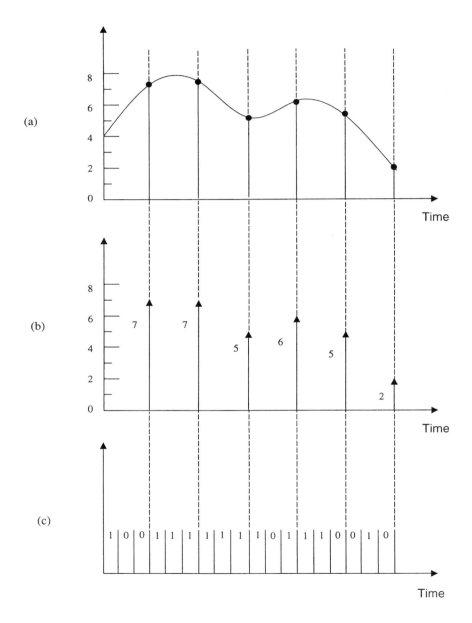

Fig. 13.2 Analog-to-digital signal conversion. (a) Analog signal, (b) sampled signal, (c) digital signal.

Let us now consider a digital voice channel, which is operated at 64 kb/s. Since most optical fiber communication systems can handle 150 Mb/s bit-rate, it is possible to transmit several digital coded signals simultaneously in the same fiber channel, namely by using either *time-division multiplexing* (TDM) or *wavelength division multiplexing* (WDM) techniques. In other words, either multisignals can be transmitted by interleaving the channel signals to form a composite bit stream for TDM transmission or the channel signals can be spaced along the wavelength (or frequency) domain for WDM transmission, as illustrated in Fig. 13.3.

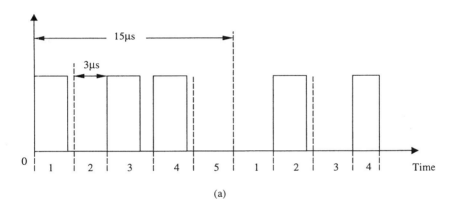

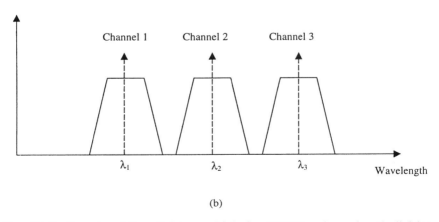

Fig. 13.3 Samples of time-division multiplexing (TDM) and wavelength-division multiplexing (WDM). (a) TDM; (b) WDM.

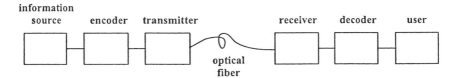

Fig. 13.4 Block-box representation of fiber-optic communication.

It is apparent that fiber-optic communications differ from other communication technologies. Besides using the light wave, it is the higher carrier frequency typically in the range of 100 Thz. In contrast with the microwave communication, the carrier frequency is in the order of 10 Ghz. Thus one would expect fiber-optic communication to be at least 10^5 times the transmission capacity than the microwave. By taking a conservative estimation, fiber-optic communication has the potential of carrying about 1 Tb/s bit-rate. In view of the current state-of-the-art optical communication systems, which are operating at about 1 Gb/s bit-rate, it is trivial to understand that fiber-optic communication would have great potential for further improvement.

To summarize our discussion, a block-box diagram fiber-optic communication is shown in Fig. 13.4, in which analog signals, from this information source, are first encoded in digital form. By converting the digital signals in light pulses, the information can be transmitted through the optical fiber. At the receiving end the light pulses are then properly decoded and then transformed into analog signals, prior to sending them to the user. The basic difference, as compared with the usual communication channels, is that the information is transmitted by coded light pulses through the fiber-optic. Although fiber-optic communication can be applied in various fields of information transfer, they have been mostly developed for telecommunication usages. There are two types of telecommunications, namely *long-haul* or *short-haul*, depending on long or short distances. Long-haul communication is mostly applied to intercity communication, which has a distance ranging from 50 to 100 km or even longer. Short-haul communication is typically used for intracity or local loop traffic, which is less than 10 km or shorter. Long-haul telecommunication systems require high-capacity trunk lines, which are benefitted mostly by the use of fiber-optic light wave technologies. Short-haul telecommunication uses a single-channel light wave system. For applications the cost-effectiveness is not generally the concern.

13.2 OPTICAL FIBER STRUCTURES

The structure of a typical optical fiber is illustrated schematically in Fig. 13.5. The center of the fiber is a cylindrical core with refractive index η_1, surrounded by a layer of material, called the cladding, with a lower refractive index η_2. The light waves are restricted by the cladding within the core and propagate along the fiber. The outer jacket layers not only protect the fiber from moisture and abrasion but also provide the strength that is required practically.

The core and the cladding are made of either silicate material (e.g., glass) or plastic. Three major combinations of these two types of material are used to make optical fibers: plastic core with plastic cladding, glass core with plastic cladding, and glass core and glass cladding. A plastic core is generally made of polystyrene or polymethyl methacrylate, while a plastic cladding is typically made of silicone or Teflon. For glass cores and claddings, the silica must be extremely pure; however, very small amounts of dopants such as boron, germanium, or phosphorous may be added to

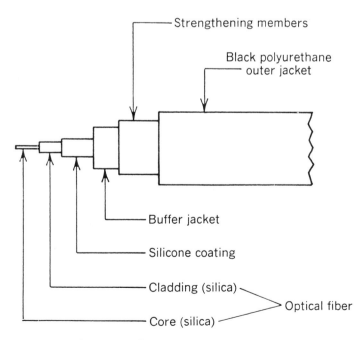

Fig. 13.5 Structure of an optical fiber.

change the refractive indices. In some claddings, boron oxide is often added to silica to form borosilicate glass.

In comparison with glass fibers, plastic fibers are flexible, inexpensive, and easy to install and connect. Furthermore, they can withstand greater stress and weigh 50% less than glass fibers. However, they do not transmit light as efficiently. Due to their considerably high losses, they are used only for short-haul communication (such as networks within buildings) and some fiber sensors. Since glass core fibers are much more widely used than plastic ones, subsequent references to fibers are to glass fibers. Glass fibers, although slightly heavier than plastic fibers, are much lighter than copper wires. For example, a 40 km long glass fiber core weighs only 1 kg, whereas a 40 km long copper wire with a 0.32 mm outer diameter weighs about 30 kg. In a fiber-optic communication network, information is transmitted by means of the propagation of electromagnetic waves along the core of the optical fiber. To minimize the transmission loss, the electromagnetic wave must be restricted within the fiber core, and not be allowed to leak into the cladding. As we shall see, the refractive index of the core η_2 must be greater than that of the cladding η_2 if such a requirement is to be met.

The diameters of the core and cladding determine many of the optical and physical characteristics of the fiber. For example, the diameter of a fiber should be large enough to allow splicing and the attachment of connectors. However, if the diameter is too large, the fiber will be too stiff to bend and will take up too much material and space. In practice, the diameter of fiber cores ranges from 5 to 500 μm, and outer diameters of fiber claddings vary from 100 to 700 μm. To keep the light wave within the fiber core, the cladding must have a minimum thickness of one or two wavelengths of the light transmitted. The protective jackets may add as much as 100 μm to the fiber's total diameter.

Optical fibers with the minimum core diameter have the advantage of having a much broader bandwidth. However, their small size may cause severe handling and connection problems. In fact, for any given fiber, tight tolerances are necessary, because even a slight variation in the dimensions can cause significant changes in optical characteristics. The tolerance for the core diameter is typically about ± 2 μm.

Although ordinary glass is brittle and easily broken or cracked, glass fibers usually have high tensile strength and are able to withstand hard pulling or stretching. The toughest fibers are as strong as stainless steel wires of the same diameter, and have the same tensile strength as copper wires with twice the diameter. For example, 1 km lengths of these fibers have withstood pulling forces of more than 500,000 lb/in.^{2} before breaking. A 10 m long fiber can be stretched by 50 cm and still spring back to its original shape, and a fiber with 400 diameter can be bent into a circle with a radius as small as 2 cm.

In order to produce fibers with such tenacity, manufacturers have to keep the glass core and cladding free from microscopic cracks on the surface or flaws in the interior. When a fiber is under stress, it can break at any one of these flaws. Flaws can develop during or after manufacturing. Even a tiny particle of dust or a soft piece of Teflon can lead to a fatal scratch on the surface of the fiber core or cladding. To prevent such abrasion, manufacturers coat the fiber with a protective plastic (or organic polymer) jacket immediately after the fiber has been drawn. This jacket also protects the fiber surface from moisture and cushions the fiber when it is pressed against irregular surfaces. The cushioning reduces the effect of small random bends (i.e., so-called microbends) which would otherwise cause transmission losses. Moreover, the jacket compensates for some of the contractions and expansions caused by temperature variations.

Although a single fiber is used in some applications, a number of fibers can be tightly packed to form a fiber cable. The cable is optimally designed such that it has little or no stress among fibers themselves.

13.3 FIBER-OPTIC TRANSMISSION

Optical fibers can be considered as cylindrical waveguides. There are two boundaries to be considered in an optical fiber: one at the core-cladding interface, and the other at the cladding-air (or cladding-jacket) interface, as shown in Fig. 13.6 in which we assume that the core has a higher

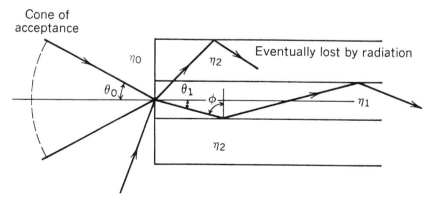

Fig. 13.6 The acceptance angle for an optical fiber. φ Is the incident angle at the core-cladding boundary.

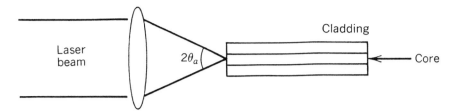

Fig. 13.7 Coupling of light into an optical fiber.

refractive index than the cladding. This figure shows how light rays are coupled into and are kept within the optical fiber. If the incident angle of a light ray at the end of the fiber core is less than the angle of acceptance, as given by

$$\theta_a = \sin^{-1}\sqrt{\frac{\eta_1^2 - \eta_2^2}{\eta_0}} \tag{13.7}$$

then the light ray will be totally reflected at the core-cladding boundary and it will be guided along the fiber case. Light rays at greater incident angles may be guided by the cladding. Since the loss in the cladding material is usually much higher than that of the core, the light rays guided along the cladding layer will eventually diminish due to absorption. Therefore, information can be transmitted only along the core of an optical fiber.

In order to efficiently couple light rays into an optical fiber, it is necessary to generate a cone of light that matches the acceptance angle of the fiber. This can be achieved by using a condenser lens with a suitable numerical aperture, as shown in Fig. 13.7. When small-sized semiconductor lasers or light-emitting diodes are used, they can also be put in close proximity to the fiber, as depicted in Fig. 13.8. In this direct coupling scheme, the light source and the fiber are generally linked by index-matching liquid

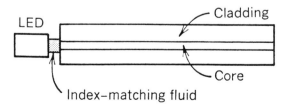

Fig. 13.8 Coupling of light with a light-emitting diode (LED).

Communication with Fiber Optics

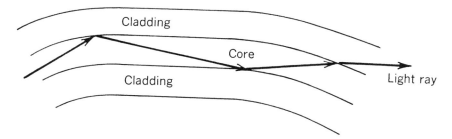

Fig. 13.9 Attenuation caused by bending.

to reduce the losses. The coupling efficiency depends on the degree of matching between the numerical apertures of the light source and the fiber.

After the light has become properly coupled into the fiber, two physical mechanisms take place during the transmission: *attenuation and dispersion*. Attenuation is the reduction in light intensity as light propagates through the fiber, which is mainly due to the chemical impurities in the fiber core. If the absorption resonance of these impurities is within the range of optical frequencies being transmitted, there will be considerable attenuation through absorption. A tremendous research and development effort has been made to reduce the amounts of these impurities. At present, optical fibers with an attenuation of less than 0.5 dB/km have been made, and an attenuation of 3 dB/km has become standard for fiber communication networks.

There are, however, two other sources of attenuation, both of which are related to the effect of bending, as illustrated in Figs. 13.9 and 13.10. These diagrams depict the escape of light rays from the fiber core, which are just within the critical angle before bending or microbending occur. By "bending" we mean the actual bending of the whole fiber, but a

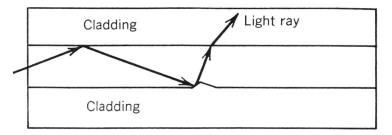

Fig. 13.10 Attenuation caused by microbending.

microbend is in fact a slight kink at the boundary between the core and the cladding. Notice that a considerable amount of light will leak when the bending (or microbend) is sharp. In this context, a sharp bend is one having a radius of about 50 mm or less. Such bends can lead to a loss of a few decibels per kilometer.

Dispersion of light transmission in the optical fiber has two general causes. The first one is called material dispersion, and results in the separation of a complex electromagnetic wave into its various frequency components. It is the result of the dependence of the velocity of light on the wavelength, or in other words, the dependence of the refractive index of the fiber material on the wavelength $\eta(\lambda)$, a circumstance that is similar to the chromatic aberration of a lens.

The second type of dispersion is called *intermodal dispersion*, which occurs even with purely monochromatic light waves, and is caused by different paths of light rays propagate the fiber. This can be easily understood from the fact that two light rays are traveling along the fiber: One travels along the optical axis and the other zig-zags up and down at the critical angle. Since these two light rays travel different distances along the fiber core to arrive at the same destination, the relative path difference between these two light rays can be shown as

$$\frac{\Delta l}{l} = \frac{\eta_1 - \eta_2}{\eta_2} \tag{13.8}$$

where l is the distance between the light ray. The intermodal dispersion within the optical fiber in terms of the difference in the periods of time that are taken by these two light rays can be shown as

$$\Delta T - \left(\frac{\eta_1 - \eta_2}{\eta_2}\right)\frac{L\eta_1}{c} = (\eta_1 - \eta_2)\left(\frac{\eta_1}{\eta_2}\right)\frac{L}{c} \tag{13.9}$$

where L is the length of the optical fiber and c is the velocity of light in vacuum.

In the preceding discussion, geometrical optics have been used to explain the propagation of light rays along optical fibers. However, the transmission of light along optical fibers can be better characterized by using electromagnetic wave theory. Similar to the laser oscillation modes, the propagation of light waves in an optical fiber is not arbitrary, but has to take some discrete propagation modes which are determined by the boundary conditions. For simplicity, we first assume that the optical fiber behaves as a waveguide with perfect conducting walls. Under such circumstances, the electric field at the waveguide boundary must be zero. An example of the propagation mode that meets such a requirement is illustrated in Fig. 13.11.

Communication with Fiber Optics 307

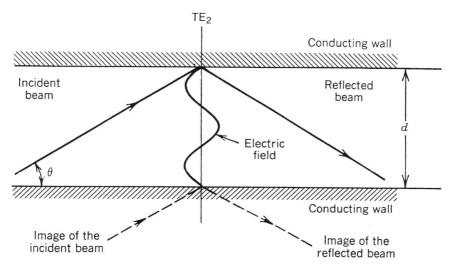

Fig. 13.11 A conducting waveguide. TE$_2$, transversal electric field of order 2.

It should be noted that the electric field across the waveguide consists of interference patterns between the incident and reflected beams. To satisfy the boundary conditions (i.e., $E_b = 0$), the diameter of the waveguide must intersect an integral number of half-wavelengths of the incident light. This can be expressed as

$$\sin\theta = \frac{(n+1)\lambda}{2d\eta} \qquad n = 0, 1, 2, 3, \ldots \tag{13.10}$$

where θ is the incident angle, d is the separation of the conducting walls, η is the index of refraction, and n is the propagation mode. Thus, we can see that only certain values of the incident angles allow propagation along the waveguide; otherwise the boundary condition would be violated.

An exact solution of the modal description for optical fibers may be obtained by solving the Maxwell's equation subject to the boundary conditions. The detailed derivations are beyond the scope of this section, and interested readers can find them in a number of texts (e.g., Refs. 5–7). Nevertheless, a similar solution for the conducting waveguide can be applied to optical fibers. The solution shows that only a number of specific angular directions of the incident beam can be allowed to propagate along the fiber. In fact each of the allowed beam directions corresponds to a different mode of wave propagation in the fiber.

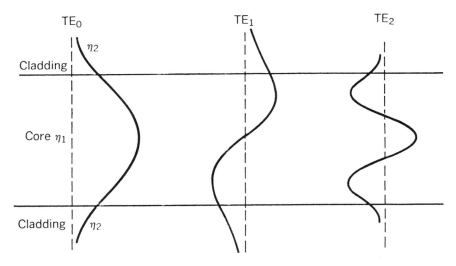

Fig. 13.12 Distribution of transversal electric field across the core-cladding region for different modes.

To discuss further the modal description in optical fibers, we can now equivalently exam the model field in a planar dielectric slab waveguide, as shown in Fig. 13.12. The dielectric slab corresponds to core that is sandwiched by two dielectric cladding layers. This simple planar waveguide helps us to understand the light-wave propagation in an optical fiber, which is difficult to deal with due to its cylindrical shape. A set of low-order transversal electric guided modes obtained from the solutions of Maxwell's equations are shown in Fig. 13.12. The order of the mode is equal to the number of transversal field minima (zero field points) across the waveguide, and is also related to the angle of the incident beam as given by Eq. (13.10). It can be seen that the steeper the incident angle, the higher the order of the guided wave mode. Unlike the perfect conducting waveguide, the electric fields of the guided modes are not completely confined within the central core, but go beyond the core-cladding boundary and decay very rapidly in the cladding layer. For lower-order guided modes, the electric fields are more concentrated near the center of the core. However, for higher-order modes, the electric field distribution spreads the core-cladding boundary and penetrates further into the cladding layer. In addition to the guided modes, there exists also *leaky modes* in optical fibers. These leaky modes are only partially confined in the core region of the fiber. As they propagate along the fiber their intensity attenuates continuously by radiating out from the fiber core. It can be shown that the upper and lower bounds of the propagation constant β for the existence of such leaky modes is given by

$$\eta_2 k < \beta < \eta_1 k \tag{13.11}$$

where $k = 2\pi/\lambda$. Most of leaky modes disappear after a few centimeters along the fiber due to significant radiation losses. However, under certain circumstances a few leaky modes may persist in the fiber for over a 100 m.

13.4 TYPES OF OPTICAL FIBERS

In general, optical fibers can be divided into two types. They are *single-mode* and *multimode* fibers. For single-mode fibers, only one guided mode is permitted to propagate along the fiber core. This transmitting mode has the lowest order ($n = 0$) and corresponds to the transversal electric field profile that has no minima across the fiber cross section. All the other orders ($n \geq 1$) have an incident angle at the core-cladding boundary that is smaller than the critical angle and, consequently, quickly leak into the cladding layer and are absorbed. By referring to Eq. (13.10), the condition for single-mode operation is

$$\theta \approx \sin\theta < \frac{\lambda}{d} \tag{13.12}$$

where d is the diameter of the fiber core. In terms of critical angle at the core-cladding boundary, the single-mode operation condition can be expressed as

$$\theta = \frac{\pi}{2} - \phi_c < \frac{\lambda}{d} \tag{13.13}$$

or

$$d < \frac{\lambda}{\pi/2 - \phi_c} \tag{13.14}$$

This implies that, in order to achieve single-mode operation, the fiber core can only have a very small diameter, ranging from 3 to 10 μm depending on the wavelength of light to be transmitted.

The dispersion in a single-mode fiber is primarily due to absorption by the material of the core. Therefore, single-mode fibers are the most suitable media for *long-distance communication* and are capable of handling very broad band signals. Since the core diameter of a single-mode fiber is generally very small, launching light into the fiber is rather difficult, and keeping the insertion of light constant over a long period of time is even more

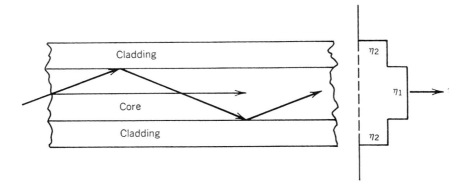

Fig. 13.13 Step-index fiber.

challenging. Fortunately, these problems can be solved with the advances in optoelectronic technology, for which single-mode fibers have been deployed in various high-capacity communications networks.

In a multimode fiber, a number of guided modes (from two to a few thousand) are transmitted simultaneously. There are two types of multimode fiber: *step-index fibers* and *graded-index fibers*. A step-index fiber is, in fact, the type of fiber that we have been discussing so far. In this type of fiber, the refractive index is constant within the core, but changes abruptly at the core-cladding interface, as shown in Fig. 13.13. Although they have a simple structure, this type of fiber suffers from severe intermodal dispersion. Ray diagrams for two different propagation modes (represented by an axial ray and a reflected ray) are shown in Fig. 13.13. One percent of optical path difference between these two light rays is common, which may lead to a time delay of 40 ns between the two modes after they have propagated 1 km along the fiber. This makes it impossible to transmit information at a data rate of 20 Mbit/s over a 1 km communication link using step-index fibers.

In contrast, a graded-index fiber does not have a constant refractive index within its core, but has a decreasing index $\eta(r)$ as a function of radial distance. As shown in Fig. 13.14, the refractive index varies from a maximum value η_1 at the center of the core to a constant value η_2 at the core-cladding boundary. The profile of the refractive index can be generally described by the expression

$$\eta(r) = \begin{cases} \eta_1 \left[1 - 2\Delta\left(\frac{r}{a}\right)^\alpha\right]^{1/2} & r < a \\ \eta_2 & r \geq a \end{cases} \quad (13.15)$$

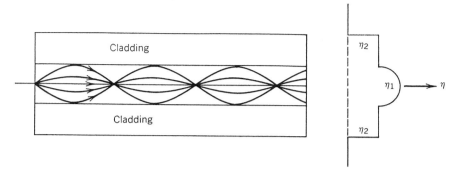

Fig. 13.14 Graded-index fiber.

where a is the radius of the core, α is the profile parameter, r is the radial distance, and

$$\Delta = \frac{\eta_1^2 - \eta_2^2}{2\eta_1^2} \approx \frac{\eta_1 - \eta_2}{\eta_1} \tag{13.16}$$

is known as the relative *refractive index difference*. The refractive index profile varies with the profile parameter α as shown in Fig. 13.15. It can be seen

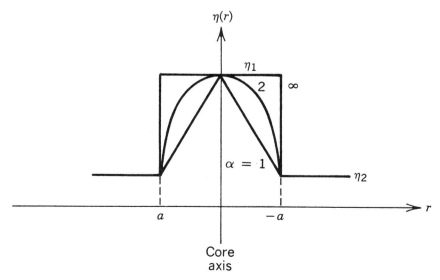

Fig. 13.15 Refractive-index profiles for graded fiber.

that when $\alpha = \infty$, the profile actually represent that of the step-index fiber, $\alpha = 2$ corresponds to a parabolic profile, and $\alpha = 1$ illustrates a triangle profile.

The ray paths in a graded-index fiber core are curved, as illustrated in Fig. 13.14. Since the inner rays travel through a higher refractive index and the outer rays travel through a lower index, the difference in the propagation time period for the outer and inner rays to be transmitted through the same fiber length can be made very small. In practice, such a time difference ΔT can be reduced to less than 1 ns/km for a core diameter of about 50 μm. In other words, the graded-index fibers have much less intermodal dispersions than the step-index fibers. Since launching the light into a 50 μm core is fairly straightforward, it is possible to use butt-joint coupling with a semiconductor laser or a light emitting diode, as shown in Fig. 13.8.

13.5 FIBER OPTIC COMMUNICATIONS

Owing to the successful reduction in optical fiber losses, more than 10 million km of fiber has been installed worldwide in various optical networks. Today bit rates for long-haul fiber links typically range between 400 and 600 Mbit/s and in some cases up to 4 Gbit/s, and are beyond the reach of copper cable communication networks. It is expected that fiber-optic communication will play a very important role in the development of the *information superhighway*, which will have a profound impact on many aspects of our society.

A typical fiber optic communication system may be presented in Fig. 13.16, in which light sources of dimensionally compatibility with the fiber core can be used. In practice, LEDs and laser diodes can be used, since their output light can be easily modulated. The electrical input signal can be either analog or digital form, although digital signals are utilized in most telecommunications networks. The transmitter circuitry converts the electrical input signals into optical signals by simply varying the current flow through the light source. Output optical signals from the channels are then multiplexed with a multiplexer prior coupled into the optical cable. Connectors and couplers are utilized along the optical cable to extend the fiber link. This has the advantages for restoring the connection after repair, maintenance, and for switching between channels. At the end of the optical cable, the transmitted signals are first demultiplexed and then detected by an array of photo cells, for which semiconductor PIN and avalanche photodiodes are commonly used. The array of detected signals are then amplified and reshaped before they were delivered to the user (e.g., telephone or computer terminals).

Communication with Fiber Optics

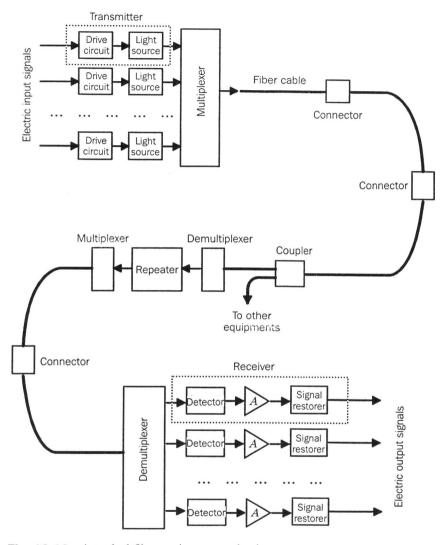

Fig. 13.16 A typical fiber-optic communication system.

Optical signal propagation along the fiber will be progressively attenuated and distorted. This is primarily due to scattering, absorption, and dispersion along the fiber. Repeaters are used along the transmission line to amplify and reshape the signals before it becomes significantly deteriorated. An optical repeater consists of a receiver and a transmitter can be used to rejuvenate the signals. In other words, this repeater detects the signal, amplifies it, reshapes it, and then sends it down to the optical fiber link. Thus we see that aside from light sources and photodetectors, long-distance fiber-optic communication requires transmitters, receivers, repeaters, connectors, couplers, multiplexes, and demultiplexers. These are the key components needed for a general optical communication system.

To take advantage of the transmission capacity of an optical fiber, signals from multichannels can be transmitted through a single fiber using multiplexing techniques. There are generally two types of multiplexing methods: *time-division multiplexing* (TDM) and the *wavelength-division multiplexing* (WDM).

Although in practice time-division multiplexing and demultiplexing systems may be complicated, their operational principle is rather simple and straightforward, as can be seen by the two-channel system illustrated in Fig. 13.17. The signals from different channels are first encoded into series of short pulses and then launched into the fiber cable with a time delay δt. As can be seen, the multiplexer can be realized with a mirror and a beam splitter. It should be mentioned that the light waves launched into the fiber are linearly polarized. At the end of the fiber link, the transmitted signal is demultiplexed by a demultiplexer composed by an electrooptic modulator and a polarization beam splitter. This can be accomplished by rotating the polarizer of channel 1 clockwise, and the polarizer of channel 2 counterclockwise. The rotated polarized beams are then separated by a polarized beam splitter and then directed to two photodetectors.

Since the light source used for fiber-optic communication has a narrow spectral width, TDM utilizes only a very small transmission bandwidth. It is apparent that a drastic increase of the transmission capacity of a fiber link can be achieved by transmitting the signals using different wavelengths from different light sources. This is the basis of WDM. Conceptually, WDM scheme is the same as the *frequency-division multiplexing* used in the microwave radio and satellite communication systems. A block diagram of the WDM system is given in Fig. 13.18, in which we see that the signal in each wavelength channel has been multiplexed via a TDM scheme.

Notice that angular dispersive elements, such as prisms and gratings, can be used as multiplexers in the WDM systems. In other words, light rays from different channels (and hence at different wavelengths) can be directed by a prism or grating at different incident angles. At the output of the multiplexer, these light rays become collinear and can be easily launched simultaneously into an optical fiber. Needless to say that a wavelength-division demultiplexer works in exactly the reverse fashion as a WDM, by which it directs the light beams of different wavelengths (from the fiber) into their respective channels.

Communication with Fiber Optics

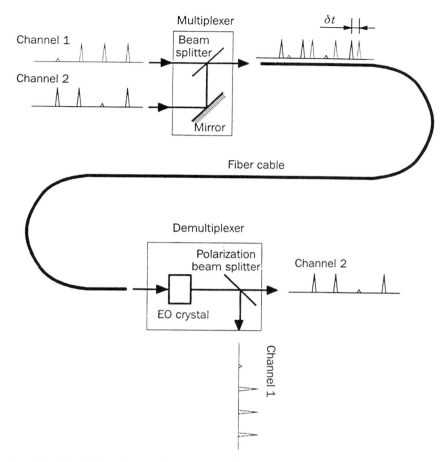

Fig. 13.17 TDM fiber-optic communication.

13.6 REMARKS

We emphasize again that the important aspects of using single-mode fiber for information transmission must be the wide bandwidth and low loss. Currently, fabrication of optical fibers has become a relatively mature technology, waiting for us to fully exploit its capability for various possible applications. Even with state-of-the-art optoelectronics-integrated circuit (OEIC) technologies, the potential capability of using optical fiber as applied to information-transmission is still far below expectations. In other words, the future development on fiber-optic communication depends upon

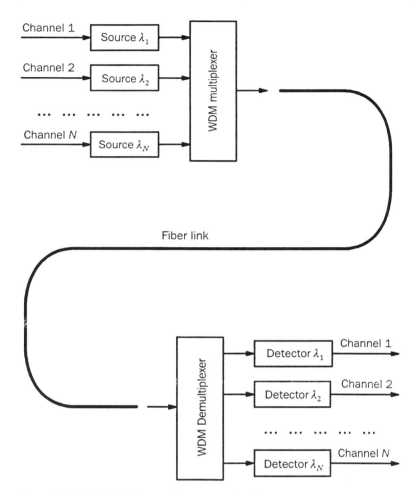

Fig. 13.18 WDM fiber-optic communication.

the successful investigation on high-speed and high-density WDM technology, for which operation beyond 20 Gb/s bit-rates is feasible. Added with the recent development of OEIC chips, it is certain that the OEIC technology would play a crucial role for developing high-density and high-speed optical communication systems. Nevertheless, coherent optical fiber communication will be the technology that will certainly be used for multichannel WDM applications. In other words, the future long-haul information-transmission communication will rely on the development of OEIC coherent transmitters and receivers.

One of the new surge applications of fiber-optic must be the subscriber-loop communicaton which includes the loop-feeder and distributed communication links. The loop-feeder loop network will be typically used for longer 10 km range, and the distributed subscriber loop will be used in more localized communication between 1 to 3 km long. Since absorption loss and dispersion are the main issues for long-haul communication, these unfavorable aspects would not be a major concern for subscriber loop applications. Instead the once-for-all installation, frequent upgrading, and maintenance are major issues for the localized communication. We anticipate that activities of near future fiber-optic communication would be more concentrated in localized transmission, for example, home subscriber deployment. Of course, the realization of this new surge of application will heavily depend upon the dramatic development on WDM and OEIC technologies.

REFERENCES

1. A. Hasegawa and F. Tappert, "Transmission of stationary nonlinear optical pulses in dispersive dielectric fibers. I. anomalous dispersion," *Appl. Phys. Lett.*, vol. 23, p. 142, 1973.
2. L. F. Mollenauer and K. Smith, "Demonstration of soliton transmission over more than 4000 km in fiber with loss periodically compensated by Raman gain," *Opt. Lett.*, vol. 13, pp. 675–677, 1988.
3. E. Yamada, K. Suzuki, and M. Nakazawa, "Gbit/s single-pass soliton transmission over 1000 km," *Electron. Lett.*, vol. 27, p. 1289, 1991.
4. L. F. Mollenauer, B. M. Nyman, M. J. Neubelt, G. Raybon, and S. G. Evangelides, "Demonstration of soliton transmission at 2.4 Gbit/s over 12000 km," *Electron. Lett.*, vol. 27, p. 178, 1991.
5. G. Keiser, *Optical Fiber Communications*, 2nd ed., McGraw-Hill, New York, 1991.
6. W. Etten and J. Plaats, *Fundamentals of Optical Fiber Communications*, Prentice Hall, Englewood Cliffs, NJ, 1991.
7. G. P. Agrawal, *Fiber-Optic Communication Systems*, 2nd ed., Wiley-Interscience New York, 1997.
8. F. T. S. Yu and X. Yang, *Introduction to Optical Engineering*, Cambridge University Press, Cambridge, UK, 1997.

Appendix A
Linear Difference Equation with Constant Coefficients

An equation of the form

$$U(x+n) + P_{n-1}U(x+n-1) + \cdots + P_1 U(x+1) + P_0 U(x) = f(x) \tag{A.1}$$

where P_i takes on the values $i = 0, 1, \ldots, n-1$ and $f(x)$ is a function of x, or constant, is called a *linear* difference equation. If $f(x) = 0$, then the equation is called a *homogeneous linear* difference equation. If $f(x) = 0$ and all the coefficients are constants, then the equation is known as a homogeneous linear difference equation with constant coefficients.

It is noted that linear difference equations with variable coefficients are extremely difficult to solve. However, if the variable coefficients are under very severe restrictions, a linear difference equation may be solvable. Linear difference equations with constant coefficients are generally solvable.

Now let us consider a linear homogeneous difference equation with constant coefficients:

$$U(x+n) + A_{n-1}U(x+n-1) + \cdots + A_1 U(x+1) + A_0 U(x) = 0, \tag{A.2}$$

where A_i, $i = 0, 1, \ldots, n-1$, are constants. It is noted that Eq. (A.2) has n *particular* solutions. These particular solutions can be found by letting

$$U(x) = a^x \tag{A.3}$$

where $a \neq 0$.

By substituting Eq. (A.3) in Eq. (A.2), we have

$$a^{x+n} + A_{n-1}a^{x+n-1} + \cdots + A_1 a^{x+1} A_0 a^x = 0 \tag{A.4}$$

or
$$a^x(a^n + A_{n-1}a^{n-1} + \cdots + A_0) = 0 \tag{A.5}$$

Therefore, if a is a solution of Eq. (A.5), then it is necessary that
$$a^n + A_{n-1}a^{n-1} + \cdots + A_0 = 0 \tag{A.6}$$

where Eq. (A.6) is also known as a *characteristic* equation. Conversely, if $a_1, a_2, \ldots, a_n$ are n distinct solutions of Eq. (A.6), then $a_1^x, a_2^x, \ldots, a_n^x$ are independent solutions of the homogeneous equation (A.2). Furthermore, the linear combination

$$U(x) = \sum_{k=1}^{n} C_k a_k^x \tag{A.7}$$

is also a solution, where C_k are arbitrary constants. We can prove that Eq. (A.7) is also the general solution of Eq. (A.2). It is noted that, for the case of multiple roots of a_i of degree m,

$$U(x) = a_i^x, xa_i^x, \ldots, x^{m-1}a_i^x \tag{A.8}$$

is also an independent solution.

Appendix B
Solution of the *a priori* Probabilities of Eqs. (5.37) and (5.38)

Let $p_n(t)$ be the probability of n molecules arriving at a trapdoor at time t, and let λ be the average rate of molecules arriving at the trapdoor. Then for $n=0$ the well-known Kolmogorov differential equation reduces to

$$\frac{\partial p(t)}{\partial t} = -\lambda p(t) \tag{B.1}$$

with the initial condition

$$p(0) = 0 \tag{B.2}$$

The solution therefore is

$$p(t) = e^{-\lambda t} \tag{B.3}$$

where $p(t)$ is the probability that no molecules will arrive at the trapdoor.

It is noted that $\bar{n} = \lambda \, \Delta t$, the average number of molecules arriving at the trapdoor per Δt. Thus the *a priori* probability of p_2 is

$$p_2 = p(t)|_{t=\Delta t} = e^{-\bar{n}} \tag{B.4}$$

Hence

$$p_1 = 1 - p_2 = 1 - e^{-\bar{n}} \tag{B.5}$$

Appendix C
Probability Energy Distribution

Let a physical measuring device or apparatus be maintained at a constant temperature of T degrees Kelvin. This temperature remains practically constant as long as a very small quantity of heat (as compared with the total energy E_0 of the device) is involved. Let the entropy equation of the device be

$$S = k \ln N_0 \tag{C.1}$$

where k is Boltzmann's constant, and N_0 is the amount of complexity of the device. Then we have

$$\frac{d \ln N_0(E_0)}{dE_0} = \frac{1}{kT} \tag{C.2}$$

It is apparent that the solution of Eq. (C.2) is

$$N_0(E_0) = Ce^{E_0/kt} \tag{C.3}$$

We now introduce a small quantity of heat (energy ΔE) into this device. It is assumed that the excess energy is very small by comparison with E_0, so that the system temperature T is not altered appreciably. However, this excess energy, although very small, causes certain small energy fluctuations within the device, such as

$$E_0 = E - \Delta E \tag{C.4}$$

Thus from Eq. (C.3) we have

$$N_0(E_0) = Ke^{-\Delta E/kT} \tag{C.5}$$

where $K = Ce^{E_0/kT}$.

Since the most probable energy distribution is the one for which the complexity N of the device is maximum, we have

$$N = N_0(E_0) \Delta N (\Delta E) = \Delta N (\Delta E) \, Ke^{-\Delta E/kT} \tag{C.6}$$

This equation shows that in the physical device each of the small changes in complexity ΔN corresponds to a certain energy ΔE involved. Thus the probability distribution of energy can be written

$$p(\Delta E) = K \exp\left(-\frac{\Delta E}{kT}\right) \tag{C.7}$$

where K is an arbitrary constant. It is also noted that Eq. (C.7) is essentially the well-known Gibbs' formula in statistical thermodynamics.

Index

Accuracy, definition of, 145
  in observation, 145–152
Adaptive neural network, 260
Additive Gaussian channel, 24
Additive noise, 21, 57, 195
Additivity, definition of, 42
Ambiguity function, 64
Amplitude filter, 179, 188
Amplitude spectrum, 44
Analog-to-digital conversion, 298
Analytic, 53
  continuation, 176–178
  function theory, 176
  signal, 53
Angle of acceptance, 304
Artificial intelligence, 284
Area of ambiguity, 64
Associative memory, 259
Average cost in statistical
    detection, 58

Ban-limited, definition of, 49
  channel, 26, 195
  signal, 53
Band-pass filter, 45–46
  cut-off frequency, 45
Bandwidth, definition of, 44–45
  analysis, 45–52
  effective, 179, 206
  norminal, 46

[Bandwidth]
  required, 200
  spatial, 76, 179
  system, 76
Bayesian decision process, 57–60
Bayes' theorem, 57, 289
Bessel function, 157
Best condition, 220
Binary, 5
  code, 11
  erasure channel, 20
  signal, 20, 57
  source, 20, 105
  symmetric channel, 20
  unit, 5
Binomial, distribution, 135
  expansion, 40
Biorthogonality, 236
  constraints, 241
Blackbody radiation, 112, 199, 212
Blurred photographic image, 186
  of restoration, 186–192
Boltzmann's constant, 29, 100
Boolean function, 273
Boundedness, 175–179
Brillouin, L., 98, 107
Brownian motion, 113

Carnot's principle, 95

Cascaded channels, 182–184
Celsius, A., 96
Celsius temperature, 96
Channel capacity, of classical
    limit, 18–22, 24–30, 81,
    201, 211
  of narrow band photon
    channel, 207–208
  of quantum limit, 201
Channel, (see Communication
    channel)
Characteristic, diameter,
    148–149
  distance, 148, 154
Chirp-Z transform, 281
Clausius, R., 96
Coding, 11
  binary, 11, 90–91
  efficiency, 92
  optimum, 81
  principle, 75–92
  ternary, 90
  theorem, 11
  theory, 70
  uniquely decodable, 70
Coherence, 217–224
  degrees, 218, 221
  exploitation, 226
  illumination, 226
  image enhancement, 165,
    178–180, 230
  length, 219
  longitudinal, 223
  mutual coherence, 217
  partial coherence, 219
  processing, 72–73, 133–135,
    186–192, 224–232
  requirement for processing,
    230–238
  self-coherence, 218
  source encoding, 229

[Coherence]
  spatial coherence, 219,
    221–223
  temporal, 222, 223
  theory, 217
  transversal, 223
  visibility, 218
  Young's experiment, 219
Communication channel, 3–31
  additive noise, 22, 30, 195
  band limited, 30, 195
  binary symmetric, 21
  capacity, 16–30, 78–80,
    144–145, 195, 201–205,
    207–208
  continuous, 15, 22–23, 195
  definition of, 15, 18, 23, 30
  deterministic, 6, 12
  discrete, 15–21
  doubly uniform, 18
  encoding, 2
  equivocation of, 13, 181, 183,
    197, 205
  erasure, 20
  finite memory, 4
  matching a code, 89–92
  memory, 4, 16
  memoryless, 16–29
  narrow band, 29, 195
  noiseless, 6, 12, 21
  nonstationary, 16
  photon, 195–205
  one-way, 15–16
  optical, 75–91, 180, 195
  quantum effect, 195–215
  spatial, 75–93, 156, 180
  stationary, 15, 21, 195
  symmetric, 19–21
  time discrete, 22
  two-way, 16
  uniform, 19

Index 327

[Communication channel]
  wide band, 201
Complex amplitude processing,
    226, 228
Complex degree of coherence,
    218
Complex spatial filter, 186
Composite filters, 263
  simulated annealing, 266
  synthetic discriminant
    function, 283
Computing, 271–293
  artificial intelligence, 284
  expert system, 284
  logic-based, 271
  matrix-matrix processing,
    282
  matrix-vector multiplication,
    278
  perfect shuffle, 276
  shuffling, 275
  systolic processing, 279
Conservation of energy, law of,
    95–97
Continuous channel, (see
    Communication channel)
Convolution integral, 41, 120
Correlation detection, 230, 252
Correlators, 251
  of joint-transform, 253
  of optical disk based, 255
  of photorefractive based, 257
  of Vander Lugt, 252
Cost, of encoding, 80
  of entropy, 100–107, 126–129,
    167, 191–198
  of information, 100–101
  of a measurement, 134
  minimum entropy, 124–131
  of an observation, 148–155
  of signal detection, 58–60

Crossbar switch, 279
Cross-correlation, 218

Data processing, 71
  coherent optical, 70–73, 133,
    165–171, 186–192, 230
Decision process, 57, 150
Decoding, 2
Defocusing, see Image,
    restoration
Degradation of energy, law of,
    95–97
  Kelvin's principle, 95
Degree of coherence, 218, 221
  spatial, 226
  temporal, 228
Degrees of freedom, 49, 158,
    209–210
  of a signal, 49–52
Degree of enhancement,
    definition of, 179
Demon, diffraction limited, 123
  efficiency, 117, 120
  exorcists, 109–131
  Maxwell's, 111–114, 118
  pressure, 118
  temperature, 108, 111
  Szilard's, 121–128
Detectability, 171–173
Detection of signal, 54–57, 55–61,
    230
Deterministic 12, 16
Difference equation, definition
    of, 319
  characteristic equation, 320
  homogeneous, 319
  linear, 319
  solution, 320
Diffraction, definition of, 35
  Fraunhofer, 36
  Fresnel, 36

[Diffraction]
  Huygens' principle, 38
  Kirchhoff's integral, 41
  limit, 179–181, 186–192
  limited demon, 124
  scalar theory, 35
Digital filtering, 180
Digital image enhancement, 179–180
Dirac dilta function, 170
Discrete channel, *see* Communication channel
Discrete linear transformation, 280
  cosmetransform, 281
Disorder, 8, 106–107
  and lack of information, 107
Distribution, binomial, 140
  Bose-Einstein, 198–199
  Gaussian, 24–33, 210
  Gibb's, 135, 193, 204, 206–210, 324
  Optimum signal power, 212–215
  Planck's, 198
Duration, 46
  norminal, 46–47
  of a signal, 46–47, 52

Effective, bandwidth, 179, 206
  temperature, 199, 209
Efficiency, coding, 92
  of demon's operation, 117, 120
  of observation, 152, 155
  thermal, 110
Einstein, A., 199
Elementary signals, 50, 53–54
Encoding, 2, 11, 75–92
Energy, compensation, 165–172, 180–185
  conservation, law of, 95–96
  degradation, law of, 95–96

[Energy]
  equipartition, law of, 104
  minimum required for observation, 138–139, 141, 201
Entropy, definition of, 8, 95–96, 181
  compensation, 115, 117–131, 138–142
  conditional, 10–12, 22–23, 180–184, 197
  decrease by demon's intervention, 115–116, 120, 124, 126
  equation, 8, 95–96, 114, 138, 156, 181
  and disorder, 8, 106–107
  of an ideal monatomic gas, 103, 124
  and information, 7–14, 99–101
  loss of information, 104–107
  noise, definition of, 12
  photon source, 166
  Shannon's definition, 5, 8
  theory, 2, 8, 95–96
  thermodynamics, 95–96
Entropy power, definition of, 25
Entropy, theory of information, 2, 7–14, 95–101, 166
Equiprobability, 9, 77, 85, 99
Equivalent noise temperature, 30, 101, 212
Equivocation, 13, 181, 197
Ergodic property, 198
Error, cost, 58
  energy, 4, 5, 160
  mean-square approximation, 61, 174
  minimum criterion, 60, 61
  momentum, 48, 160
  observation, 142–145
  position, 48, 160

[Error]
  probability, 57
  restoration, 179
  time, 48, 160
Expert system, 284

False alarm, definition of, 57
Fano, R. M., 31, 33
Figure-of-merits, 258
Filtering, 45, 55, 169, 229
  inverse, 169, 187, 229
  lowpass, 45, 230
  matched, 55, 229, 263–269
  optimum, 55–61
  phase, 179, 189
Fermions, 199
Fiber optics, 301–312
  angle of acceptance, 304
  attenuation, 305
  coupling, 305
  dispersion, 305
  graded-index, 311
  leaky modes, 308
  intermodal dispersion, 306
  microbending, 305
  multi-mode, 309
  single-mode, 309
  step-index, 310
  structure, 301
  transmission, 303
Fiber-optic communication, 301–312
  dynamic range, 297
  TDM system, 313–317
  transmission capacity, 296
  WDM system, 316
Filter function, 169, 179, 187, 229, 263–269
Finite bandwidth analysis, 44–49
First law of thermodynamics, 95–95
Fourier analysis, 43–44

Fourier domain processing, 234
Fourier spectra, 50, 188, 228, 233
Fourier transformation, 43, 49, 186, 233
  complex coefficients, 44
  definition of, 43
  and Fraunhofer diffraction, 36
  performed by lens, 70–73
  of spatially bounded function, 175
  series, 43, 49
  short-time, 233
  spectrum, 43–44, 50, 178, 179, 188
  two-dimensional, 73
  and transform pair, 43
Fraunhofer diffraction, definition of, 36
Fraunhofer, J., 36
Frequency, cut-off, 44–48, 73
  for high accuracy observation, 147–149
  sampling, 50
  saptial, 50, 60, 65, 73, 78
  spectrum, 43, 178
Fresnel, A., 36
Fresnel diffraction, definition of, 36
Fresnel-Kirchhoff theory, 39–41, 170
Fringe spacing, 227
Fundamental theorem of information, 4

Gabor, D., 52, 232
Gabor's elementary signals, 52–53
Gabor's information cell, 52–53
Gabor's perpetuum mobile of second kind, 129–131
Gabor transform, 233
Game theory, 2

Gaussian channel, 24–33, 212
Gaussian distribution, 23, 25, 209
  noise, 23–25, 28–30
  signal, 23–25, 28–30
Gaussian cosine pulse, 54, 65, 67
Gaussian sine pulse, 54, 65, 67
Gaussian window, 234, 239
Gibb, J. W., 135
Gibb's, distribution, 135, 204, 206, 209
Goal-driven chaining, 285
Global connection, 279

Hartley, R. V., 5
Hartley unit, 5
Heisenberg, 46
Heisenberg uncertainty principle, 46–48, 53, 159–162
Heteroassociation NN, 261
Holography, 169, 189
  computer generated hologram, 189
  phase filter, 189
Homogeneity, definition of, 43
Homogeneous equation, 319
Hopfield model, 260, 279
Huygens, C., 38
Huygens' principle, 38 (*see also* Fresnel-Kirchhoff theory)

Ideal monatomic gas, 103, 124
Identity theorem, 175
Illumination, coherent, 226
  incoherent, 225
Image, of blurred photography, 186–192, 229
  code, 80
  degree of enhancement, definition of, 178
  encoding, 80

[Image]
  enhancement, 178–179, 186–192, 230
  and information, 165–192
  motion, 167, 186, 230
  reconstruction, 176–177
  restoration, 178–179, 186–192, 230
  smearing, 178, 186, 229
  subtraction, 230
  uncertainty, 171–174
Impulse response, definition of, 40
  free space, 40, 61, 170
Incoherent, illumination, 225
  processing, 224
  Schwarz' inequality, 56
  superposition, 231
  uncertainty relation, 46–47, 52–54, 160–161, 171
Infinite-pole function, 187
Information, definition of, 4–7, 14, 84, 99
  average, 8, 82, 85
  Gabor's information cell, 52
  binary, 20–21, 106, 123
  capacity, 9, 18–29, 78–84, 144–145, 195–212
  channel, 3–6, 18–27, 78–84, 144, 195–212
  conditional, 7–14, 17–19
  cost of entropy compensation, 99–101, 109–131
  entropy, 2–8, 96–107, 138–142, 154–156
  and entropy theory, 95–107
  equivocation, definition of, 13, 181, 197
  fundamental theorem, 4
  gain, 114–115, 167, 175
  leakage, 180
  logarithmic measure, definition of, 5

Index 331

[Information]
  and image restoration, 178–179, 186–192, 230
  loss, 13, 75, 102–106, 185
  optical, 1, 75, 180
  measure, 4–7, 15, 181, 197
  mutual, 5–14, 181, 197
  and observation, 133–163
  quantum limit, 201, 210
  rate, 78–89
  reduction, 200
  and resolving power, 175–178
  self-information, 6–15
  source, 1–14, 75, 95, 105, 196
  symmetric property, 11–13
  theory, 1–33
  transmission, 1–33
  without illumination, 133
Interconnection networks, 275
Interconnection weight matrix (IWM), 259
Interferometry, Mach-Zehnder, 133
  Michelson, 223
  observation, 153–157
Interpattern association (IPA), 261
Invariance property, 43
Inverse filtering, 166–167, 169, 187, 229
Inversion formula, 43
  for Fourier transform, 43–44
Irreversible processes, 96–98
Isolated system, 95–96, 112, 125

Joint probability, 5–27, 182–184
Joint time-frequency, 239
Joint transform correlator (JTC), 253
  of optical disk-based, 255
  of single SLM, 254

Karhunen-Loive expansion, 26–27, 179, 185
Kelvin temperature, 96–99, 101, 122–127, 133
Kelvin's principle, 95–98
  second law of thermodynamics, 95
Kirchhoff, G., 34
Kirchhoff's integral, 34, 41 (*see also* Fresnel-Kirchhoff theory)
Kolmogorov differential equation, 321

Lagrange multipliers, 86, 205
Lagrange's identity, 168
Law of energy conservation, 95
Law of large numbers, 88
Lens, Fourier transform property of, 70–72
  optical information processing, 73, 186–192, 224–232
  resolving power, 144, 148
Letter, of English, 90–91
  binary code, 90
  information provided, 90, 91
  ternary code, 91
  probability distribution, 90
Liklihood ratio, 59
  test, 60
Linear, cascaded channels, 183–185
  difference equation, 81, 319
  distorted images, 166–167, 186
  image motion, 166, 186
Linear system, definition of, 42
  additivity property of, 41
  convolution, 41, 61–63, 66, 70
  Fresnel-Kirchhoff theory, 39–41, 70, 145
  homogeneity property, 42
  impulse response, 40

[Linear system]
  and principle of superposition, 42
  response of, 42
  time invariant, definition of, 43
  transfer function, 45, 56, 143, 167, 186–190
Linear transformation, 40, 280
  chirp, 7, 281
  discrete, 280
Linearity, property of, 31–42
Liquid crystal television (LCTV), 259, 274
Logarithmic expansion, 141
Logon, definition of, 52, 234
Low-pass filter, 45
  cut-off by, 45
  system, 45

Mach-Zehnder interferometer, 133
Macroscopic states, 197
Many simultaneous observations, 139–141
Matched filter, 55–57, 263–269
Matching a code, 89–92
Maxwell, J. C., 35
Maxwell's equation, 35
Maxwell's demon, 111–121
  cost of entropy, 110, 114
  operation, 111–121
  of pressure, 118
  of temperature, 111–121
  and uncertainty principle, 150–162
Maximum allowable exposure, 165–171
Mean-square error approximation, 61, 174
Mean-square fluctutation, 23, 61
Measure, of information, 4–7

Measurement, accuracy of, 143, 148, 150
  cost of entropy, 143–146, 151
  efficiency of, 152, 155, 162
  limitations, 143, 154–159
  as observation, 144–162
  quanta required for, 154–155
  of realiability, 149–151
  uncertainty, 161
Memory, channel, 5, 16
  information storage, 76, 107
  rate of attenuation, 107
Message, information content of, 2, 80–82
Michelson interferometer, 223
Michelson's visibility, 218
Microscope, of observation, 156–159
Microscopic states, 197
Microsignal, 197
Minimum cost of entropy compensation, 123, 138, 162
Minimum error criterion, 60
Minimum distance, 148
  and quantum effect, 156
"Miss", a, definition of, 57
Modulation shutter, 170
Momentum and energy, 48, 160
  uncertainty principle, 47, 48, 52, 160
Monatomic gas, 103
Monochromatic point source, 39
Multiplexing, 299
  time-division, 299
  wavelength-division, 299
Mutual coherence, definition of, 217
Mutual information, definition of, 5–7, 12–14, 17
  average, 12, 13, 17, 24, 28, 177, 197, 204

[Mutual information]
  channel capacity, 18, 24, 28
  conditional, 12, 15
  maximum, 18, 28
  per degree of freedom, 211
  properties, 13
  triple, 14

Natural unit, 5
Narrow-band signal power, 206
Negative entropy, definition of, 99, 113
  of compensation, 115–131
  equivalent amount of information, 100
  of information, 90–101
  of measurement, 158, 184–185
  of observation, 139–162
  traded with information, 90–101
  use by the demon, 109–126
Negentropy (see Negative entropy)
Neural networks, 359
Neuron, definition of, 359
Neyman-Pearson criterion, 61
Noise, additive, 22, 55
  coherent, 232
  entropy, 13
  Gaussian distribution, 24–29, 195
  photon, 196, 198
  power, 29, 195
  quantization, 297
  spectral density, 212
  stationary, 26
  thermal, 26–29, 101, 206
  white, 27, 56, 195
Norminal, bandwidth, 47
  duration, 47
Numerical aperture, 148
Nyquist, H., 50

Nyquist sampling interval, 50
Nyquist sampling rate, 50

Observable information, 133–163
Observation, 133–163
  accuracy, 143, 145, 150
  cost of entropy, 146, 149, 151, 154, 158
  efficiency of, 152, 155, 162
  energy required, 154, 156
  error, 143, 148
  high frequencies, 138, 148, 155
  and information, 118, 133–163
  by interference, 153–162
  irreversible process, 162
  low frequencies, 139, 146, 150, 152
  made by radiation, 135–139
  by microscope, 156–159
  Mach-Zehnder interferometer, 133
  many simultaneous, 139–142
  perturbation, 161–162
  reliability, 149, 150
  sequential, 144
  uncertainty of, 156, 159–162
  without illumination, 133–134
Optical, coding, 35, 80
  communication, 1, 80–81, 181–185, 295–317
  computing, 271–293
  correlators, 251
  fiber, 66
  information processing, 1, 73, 133, 165–171, 186–193, 224–232
  interconnections, 275
  large capacity, 255
  neural networks, 259
  observation, 133–163

[Optical]
  optical disk based correlator, 255
  pattern recognition, 251–269
  photon channel, 195–215
  spatial channel, 75–92, 180–185
  shuffling, 275
Optical disk based correlation, 255
  large capacity, 255
Optics and information, 1
Optimum, coding, 11, 62–79
  decision of observation, 137
  decision process, 60, 136
  filtering, 55–57, 180–192
  image restoration, 186
  linear filter, 61, 63
  matched filter, 56
  signal detection, 55–57
  signal power distribution, 212–215
Orthogonal expansion, 26
  Fourier, 43, 49
  Kahunen-Loéve, 26–27, 180, 185
Orthogonality, 236
  constraints, 241
Orthonounal function, 26

Paradox of observation, 133–134
Paraxial approximation, 38, 40
Partial coherent processing, 228–232
Passive, device, 167, 180
  information leakage, 180–185
  optical channel, 180
Pattern recognition, 251–269
  with composite filters, 263
  with correlation, 251
  with neural networks, 259

Pauli, W., 199
Pauli's exclusive principle, 199
Perpetual motion machine, 109–113, 129–131
  Gabor's perpetuum mobile, 129
  of second kind, 109, 113, 108
  Szilard's machine, 121
Perturbation, cost of observation, 161–162
Phase, filter, 179, 189
  quadratic, 72
  spectrum, 44
  transformation, 71
  volume, 209
Photographic recordability, 171–172
Photorefractive based correlator, 257
Photon, channel, 198–215
  counter, 198
  fluctuation, 204
  noise, 204–214
  required for observation, 141–142, 149–151
  source, 196
Physically realizability, 168–170, 173–174
Planck, M., 48
Planck's constant, 48, 103
Planck's distribution, 198
Point-Pain coherence, 229
Poisson's law of rare events, 134
Positive observation, definition of, 136
Pressure demon, 118
Probability, 2
  a posteriori, 5, 57, 181
  a priori, 5, 59, 119, 321
  binomial, 140
  composite transitional, 16–17

Index

[Probability]
  conditional, 5, 14, 22, 57, 183, 203
  energy distribution, 323–324
  equiprobability, 9, 77, 84, 99
  error, 57
  Gaussian, 23–26, 30, 195
  joint, 5–27
  quantum states, 135
  reduced sample space, 136
  theory, 2, 7
  transition, 5, 14, 18
  transition matrix, 18–21
Processing, 225
  complex amplitude, 226
  Fourier domain, 234
  in intensity, 226
  matrix-matrix, 282
  matrix-vector, 278
  systolic, 279
  by wavelet transform, 233
Processor, coherent, 226
  incoherent, 226
    source, 226
  showdown-costing, 272
  systolic, 279
Product space, definition of, 4, 6, 13
  continuous, 14
  Euclidean, 4, 6, 13
  discrete, 4, 6, 13
  triple, 6
Pulse code modulation (PCM), 297
Pulse function, 48, 54
  antisymmetrical, 54
  Gabor's elementary pulses, 54
  Gaussian, 54
  length and bandwidth, 44–47
  rectangular, 46
  symmetrical, 48

Quadratic phase variation, 72
Quanta, 134
  minimum required, 207
Quantization noise, 297
Quantized radiation, 135
  average energy of, 135–137
  median energy, 136
  most probable energy distribution, 135
  probability distribution of energy level, 135–139
Quantum effect, on information transmission, 195–215
  narrow band photon channel, 206–212
  photon channel, 198–215
  optimum signal power distribution, 212–215
Quantum state, 135
  mean, 135–136, 198, 203, 205
  median, 136
Quantum theory of light, 134

Radiation, black body, 110, 199, 212
  electromagnetic, 196
  high frequencies, 138, 141, 147–148, 155, 160
  low frequencies, 139, 148, 155, 158
  Planck's distribution, 198
  quantum radiation, 135
Realizability, 167–168
Recognition, 3, 55, 167
Recordability, 171–172
Rectangular pulse, 46
Redundancy, definition of, 92
  coding, 92
  English letters, 92–93
  spatial, 75, 143–144
  temporal, 75

Relative degree of image
    restoration, 179, 189
Reliability, definition of,
    149–150
  efficiency of, 152
  high and low reliability,
    152–153
  on observation, 149–155
Resolution beyond the classical
    limit, 175–178
Resolution cell, 52, 63, 64, 76–77,
    234
  Gabor's information cell, 52,
    234
  optical, 76–77
  on spatial encoding, 76
Resolution element (see
    Resolution cell)
Resolvable image, 167
Resolving power, 142–143, 148,
    175–178
  and information, 175–178
  of lenses, 142, 148
  and physical realizability,
    175–178
  and uncertainty of
    observation, 156, 159–162
Response of a linear system,
    42–43
Restoration and information,
    165–171, 186–192,
    228–231
Retrieval equation, 239
Reversible process, 101, 110

Sampling, definition of, 50
  degree of freedom, 49
  function, 51
  Nyquist interval, 50
  Shannon's theory, 50–52
Scalar wave theory, 37
Scalegram, 247

Scaling transform, 240
Schrödinger, E., 98
Schwarz's inequality, 56
Second law of thermodynamics,
    95–97, 100
Self coherence function, 218
Self information, definition of,
    6–14
  conditional, 7, 14
Shadow-casting logic processor,
    273
Shannon, C.E., 3, 4, 24, 33
Shannon's formula, 30, 195
  fundamental theorem, 4, 32
  sampling theory, 51–52
Short-time transform, 233
  Fourier, 233
  wavelet, 239
Signals
  ambiguity, 63
  analytic, 53–54
  antisymmetric, 53–54
  elementary, 53–54
  energy density, 198
  Gaussian, 23, 54
  phase, 181
  recovering, 61
  symmetric, 48, 53–54
  Wigner representation, 66
Simulated annealing filter, 266
Sinc function, 51–56, 167, 186
Smeared image, 165–169,
    186–192
  defocused, 186–192
  restoration, 165, 186
Source dispersion, 228
Source encoding, 229
Spatial, bandwidth, 76
  bounded, 175
  channel, 77, 84–92
  coherence, 221
  coherence function, 225

Index 337

[Spatial]
  coordinates, 48
  domain, 76
  encoding, 76–92
  filter, 166–167, 186, 266
  frequency, 48, 175–176
    spectrum, 175
  impulse response, 40, 70, 170
  redundancy, 75
Spatial domain filter, 266
Spatial light modulator (SLM), 252
Spectrogram, 247
Stational noise, 26, 56
Statistical, communication, 149
  definition of information, 5
  independent, 5, 10, 24, 82, 177, 182, 203
  signal detection, 55–61
  thermodynamics, 95–99, 100
Stirling's approximation, 82, 85
Superposition principle, 42
Symmetric pulse, 48, 54
Synthesis of complex spatial filter, 186–192, 263–265
Synthetic discriminant filter (SDF), 263
System bandwidth, 76
Systolic processor, 279
Szilard, L., 121
Szilard's machine, 121–124

Taylor expansion, 176
Target tracking, 212
Temperature, absolute, 96
  demon, 111
  effective, 199
  equivalent, 212
Temporal coherence, 223
  exploitation, 227
Temporal redundancy, 75
Ternary code, 91

Time, continuous, 22
  discrete, 22
  division, 299
  domain, 72
  duration, 44–48
  interval, 76
  invariant, 43
  Gabor's information cell, 52
  logon, 52
  maximum allowable exposure, 165
  nominal duration, 47
  time-frequency resolution, 42, 239
  uncertainty relation, 47, 52
Time-division-multiplexing (TDM), 299
Theory of heat, 111
Thermal, agitation, 30, 136
  Brownian motion, 113
  efficiency, 110
  energy, 158
  noise, 25, 30, 136, 146, 196, 207
  reliability, 149–150
  temperature, 30, 96, 122, 138, 156
Thermodynamics, laws of, 95–99
Threshold decision level, 58–61, 136–137, 165
Threshold level, 171–172
Transfer function, definition of, 45, 56
  of inverse filtering, 167–168, 179, 187
  of linear system, 45, 56
  of matched filtering, 56
Transform pair, 43
  Fourier, 43
  matrix, 280
Transition probability matrix, 18–21
Transmission capacity, 296

Transmission of information, 1–33
Transmittance, 169
  amplitude, 169
  phase, 71
  of thin lens, 71
Two's complement, 281

Uncertainty, definition of, 8
  Heisenberg's principle, 46–48, 53, 159–162
  and image restoration, 165–192
  and information, 8, 159–162
  and observation, 159–162
  relation for time and frequency, 46–48, 53, 64
  theory, 2
Uniqueness theorem, 175
Uniquely decodable code, 75
Uniform channel, 18–21
  definition of, 18
  of doubly, 18
  from input, 18
  from output, 18
  n-ary, 18–20
Units, 5
  binary, 5
  Hartley, 5
  natural, 5

Van Citter-Zernike theorem, 225
van der Waals force, 113

Vander Lugt correlator (VLC), 252
Visibility, definition of, 218

Wave equation, 156, 219
Wave front reconstruction (*see* Holography)
Wave guide, 157, 307
Wave number, 37
Wavelength-division multiplexing (WDM), 299, 316
Wavelength multiplexed filter, 257
Wavelet transform (WT), 233–249
  admissibility conditions, 240
  daughter wavelet, 234
  Mayer's wavelet, 245
  mother wavelet, 234
  optical implementation, 242
  scaling transform, 240
White noise, 13, 27
Wiener, N., 3, 4, 30
Wiener-Hopf integral equation, 62
Wigner signal representation, 66–70
Work done, 96

Young's experiment, 219, 221

Zone lens, 170
Zone plate, 170